International Conference on Advanced Technologies in Design, Mechanical and Aeronautical Engineering (ATDMAE 2017)

IOP Conference Series: Materials Science and Engineering
Volume 234

Singapore
12 – 14 July 2017

ISBN: 978-1-5108-4777-4
ISSN: 1757-8981

Printed from e-media with permission by:

Curran Associates, Inc.
57 Morehouse Lane
Red Hook, NY 12571

Some format issues inherent in the e-media version may also appear in this print version.

Printed by Curran Associates, Inc. (2017)

For permission requests, please contact the Institute of Physics
at the address below.

Institute of Physics
Dirac House, Temple Back
Bristol BS1 6BE UK

Phone: 44 1 17 929 7481
Fax: 44 1 17 920 0979

techtracking@iop.org

Additional copies of this publication are available from:

Curran Associates, Inc.
57 Morehouse Lane
Red Hook, NY 12571 USA
Phone: 845-758-0400
Fax: 845-758-2633
Email: curran@proceedings.com
Web: www.proceedings.com

Table of contents

Volume 234

2017 International Conference on Advanced Technologies in Design, Mechanical and Aeronautical Engineering (ATDMAE 2017)

12–14 July 2017, Nanyang Technological University, Singapore

Accepted papers received: 1 September 2017
Published online: 8 September 2017

Preface

011001
OPEN ACCESS
2017 International Conference on Advanced Technologies in Design, Mechanical and Aeronautical Engineering (ATDMAE 2017)

011002
OPEN ACCESS
Committee

011003
OPEN ACCESS
Peer review statement

Advanced Mechanical and Aeronautical Engineering

012001
OPEN ACCESS
Landing Distance Minimization to Prevent Overrun Accidents Using Field Theory and Stabilizing Air Traffic – A Novel Approach

R Krishna Kumar, M Navaneeth and R Shachin Shibi.....1

012002
OPEN ACCESS
Examining Mechanical Strength Characteristics of Selective Inhibition Sintered HDPE Specimens Using RSM and Desirability Approach

D. Rajamani and Balasubramanian Esakki.....8

012003
OPEN ACCESS
Computational Study of Convective Heat Transfer in Circular Pipe using Trapezoidal Ribs

V K Patel, P Sharma and K Saxena.....15

012004
OPEN ACCESS
Development of Gas Dynamic Probe for High Total Temperature Measurement in High Speed Flow

Tahzeeb Hassan Danish, Yash Mistry, K Sathiyamoorthy, J Srinivas and P Pratheesh Kumar.....21

012005
OPEN ACCESS
Fuel Effective Photonic Propulsion

N Rajalakshmi and S Srivarshini.....28

012006
OPEN ACCESS
Numerical Investigations on Aerodynamic Forces of Deformable Foils in Hovering Motions

Yong Zhao, Zhen Yin, Xiaohui Su, Jiantao Zhang and Yuanwei Cao.....34

012007
OPEN ACCESS
Combined Effect of Surface Roughness and Wake Splitter Plate on the Aerodynamic Characteristics of a Circular Cylinder

Iyer Saisanthosh, K Arunkumar, R Ajithkumar and A R Srikrishnan.....40

012008
OPEN ACCESS
Detection of Cracks in a Cantilever Beam Using Signal Processing and Strain Energy Based Model

P Mehta, A Kureshi, S Lad, N Patel and D Sharma.....46

012009
OPEN ACCESS
Development of Double Layer Microwave Absorber Using Genetic Algorithm

Abhishek Kumar, Samarjit Singh and Dharmendra Singh.....52

012010
OPEN ACCESS
Study on Stationarity of Random Load Spectrum Based on the Special Road

Huawen Yan, Weigong Zhang and Dong Wang.....58

012011
OPEN ACCESS
Research on Aeroheating of Hypersonic Reentry Vehicle Base Flow Fields

Qin Xuguo, Shui Yongtao, Wang Yonghai, Chen Gang and Li Qiang.....64

012012
OPEN ACCESS
Study on BTT Coordinated Turn Autopilot Design for Reentry Gliding Vehicle

Li Qiang, Shui Yongtao, Wang Yonghai, Wang Gang and Qin Xuguo.....70

OPEN ACCESS
Analysis of the Thermal Characteristics of Machine Tool Feed System Based on Finite Element Method

Xiaobo Mao, Kuanmin Mao, Yikang Du, Fengyun Wang and Bo Yan.....77

012014
OPEN ACCESS
An Balancing Strategy Based on SOC for Lithium-Ion Battery Pack

Peng Li.....83

012015
OPEN ACCESS
The Formation of Vortex Structures in a Screen Cylinder Wake

Azlin Mohd Azmi, Yucen Lu and Tongming Zhou.....91

012016
OPEN ACCESS
Effect of Free-Stream Turbulence Intensity on Transonic Airfoil with Shock Wave

I Lutsenko, M Serikbay, A Akiltayev, L R Rojas-Solórzano and Y Zhao.....98

012017
OPEN ACCESS
Investigating the Effects of Robot-Assisted Therapy among Children with Autism Spectrum Disorder using Bio-markers

Jaishankar Bharatharaj, Loulin Huang, Ahmed Al-Jumaily, Mohan Rajesh Elara and Chris Krägeloh.....106

Design Engineering and Product Innovation

012018
OPEN ACCESS
Wing Rib Stress Analysis and Design Optimization Using Constrained Natural Element Method

Mohamed Amine Bennaceur, Yuan-ming Xu and Hemza Layachi.....113

012019
OPEN ACCESS
Research on the Integration of Bionic Geometry Modeling and Simulation of Robot Foot Based on Characteristic Curve

G He, H Zhu, J Xu, K Gao and D Zhu.....120

012020
OPEN ACCESS
Research on the Bionics Design of Automobile Styling Based on the Form Gene

Zhao Aili and Jiang Long.....128

012021
OPEN ACCESS
Structural Topograohy Optimization for Thin-plate Based on Isogeometric Analysis

Zhihui Zou, Denglin Zhu, Lilv Lou, Huishan Liu and Gang He.....134

012022
OPEN ACCESS
Development of Modular Outdoor Furniture Product Using Lightweight Concrete for Public Parks in Surabaya

Grace Mulyono, Diana Thamrin and Antoni.....143

012023
OPEN ACCESS
Design of Backpack to Aid Elderly for the Mazu Touring Procession in Taiwan

F L Chao, Y C Huang, J Y Su, C L Sun and C C Chen.....150

012024
OPEN ACCESS
Elderly's Family Life Supplies - Innovative Chinese Checkers Game Board

Fanglin CHAO.....156

012025
OPEN ACCESS
Towards Enhanced Affective Design: Rethinking the Notion of Design

SuKyoung Kim and Youngil Cho.....162

012026
OPEN ACCESS
Kinematical Comparison Analysis on the Discus Athletes Throwing Techniques Based on Data Project

Li Junming, Zhou Jihe and Long Ting.....168

012027
OPEN ACCESS
The Study to the Theory of Human-Machine-Integration in Product Design

Min Gao and Renhe Zhou.....173

IOP Conf. Series: Materials Science and Engineering **234** (2017) 011001 doi:10.1088/1757-899X/234/1/011001

PREFACE

ATDMAE 2017 was held in Nanyang Technological University (NTU), Singapore, during Jul. 12-14, 2017. ATDMAE 2017 was organized by Hong Kong Society of Mechanical Engineers, supported by other organizations. The conference provides a useful and wide platform both for display the latest research and for exchange of research results and thoughts in Advanced Mechanical and Aeronautical Engineering, Design Engineering and Product Innovation. The participants of the conference were from almost every part of the world, with background of either academia or industry, even well-known enterprise. The success and prosperity of the conference is reflected high level of the papers received.

The proceedings are a compilation of the accepted papers and represent an interesting outcome of the conference. This book cover 2 chapters: Advanced Mechanical and Aeronautical Engineering; Design Engineering and Product Innovation.

We would like to acknowledge all of those who supported ATDMAE 2017. Each individual and institutional help were very important for the success of this conference. Especially we would like to thank the organizing committee for their valuable advices in the organization and helpful peer review of the papers.

We sincerely hope that ATDMAE 2017 will be a forum for excellent discussions that will put forward new ideas and promote collaborative researches. We are sure that the proceedings will serve as an important research source of references and the knowledge, which will lead to not only scientific and engineering progress but also other new products and processes.

Prof. Yong Zhao, Nazarbayev University, Kazakhstan
Prof. Henri Christiaans, Ulsan National Institute of Science & Technology, Korea

IOP Conf. Series: Materials Science and Engineering **234** (2017) 011002 doi:10.1088/1757-899X/234/1/011002

COMMITTEES

Conference Chairman

Prof. Yong Zhao, Nazarbayev University, Kazakhstan

Conference Committee Chairs

Prof. Cees de Bont, Hong Kong Polytechnic University, Hong Kong

Prof. Shu Chang, National University of Singapore, Singapore

Prof. Martin L. Dunn, Singapore University of Technology and Design, Singapore

Prof. Luciënne T.M. Blessing, Singapore University of Technology and Design, Singapore

Conference Program Committee Co-Chairs

Prof. Kaspar Jansen, Delft University of Technology, Netherlands

Prof. Henri Christiaans, Ulsan National Institute of Science & Technology, Korea

Prof. Xiaohui Su, Dalian University of Technology, China

Prof. Zijian Wang, University of Kansas, USA

Prof. Bin Wang, Brunel University London, UK

Prof. Tongming Zhou, The University of Western Australia, Australia

Dr. Yang Wenming, National University of Singapore, Singapore

Local Committee

Prof. Huang Weimin, Nanyang Technological University, Singapore (Chair)

Prof. Chua Leok Poh, Nanyang Technological University, Singapore

IOP Conf. Series: Materials Science and Engineering **234** (2017) 011002 doi:10.1088/1757-899X/234/1/011002

International Technical Committee

Prof. Weihua Liu ,Nanjing University of Aeronautics and Astronautics, China (Chair)

Dr. Redouane Zitoune, Clément Ader Institute of Toulouse University, France

Prof. B. T. N. Sridhar, Anna University, India

Dr. Zhangming Wu, Cardiff University, UK

Dr. Pierre Kerfriden, Cardiff University, UK

Prof. Kuei-Yuan Chan, National Taiwan University, Taiwan

Prof. K.M.Parammasivam, Madras Institute of Technology, India

Prof. S.Thanigaiarasu, Madras Institute of Technology, India

Peer review statement

All papers published in this volume of *IOP Conference Series: Materials Science and Engineering* have been peer reviewed through processes administered by the proceedings Editors. Reviews were conducted by expert referees to the professional and scientific standards expected of a proceedings journal published by IOP Publishing.

IOP Conf. Series: Materials Science and Engineering **234** (2017) 012001 doi:10.1088/1757-899X/234/1/012001

Landing Distance Minimization to Prevent Overrun Accidents Using Field Theory and Stabilizing Air Traffic – A Novel Approach

R Krishna Kumar[1,2,a], **M Navaneeth**[1,3,b] **and R Shachin Shibi**[1,4,c]

[1] Department of Electrical and Electronics Engineering, Sri Sai Ram Institute of Technology, Chennai 600 044.
[2] No:8, Visalakshi Apartments, 10th Street, SRP Colony, Chennai – 600 082, TN, India.
[3] No: 26/8B, B1, Madurakaliamman Flats, 12th street, TG Nagar, Chennai – 600 061, TN, India.
[4] No:1, Kuzhumani Main Road, MGR Garden, Woraiyur, Trichy – 620 003, TN, India.

[a]krishnakumarrdkk@gmail.com, [b]navemurali@gmail.com, [c]shachinsibi@gmail.com

Abstract. Airplane is considered to be the pinnacle of engineering as it has proven that it is possible for a manmade object to fly. Before its invention, flying was just a dream for mankind. In such an esteemed domain, landing is the most challenging part and it is where a large number of accidents occur, especially due to overrun. As the name suggests, overrun accidents occur due to insufficient runway length. In the present study, the concept of planar electromagnetic fields is incorporated to minimize the landing distance of an aircraft, thus preventing the overrun accidents. As a result, unexpected losses can be avoided. In addition to this, the stability of air traffic control can be perpetuated and the fuel consumed during loitering time can be diminished.

1. Introduction

One of the major problems associated with aircrafts is the number of accidents due to over-run. According to the NTSB, 379 of 1332 runway accidents (1995 and 2007) were due to overruns causing 680 fatalities [1]. It is seen that 35% of the accidents have occurred due to landing overrun (LDOR), which is a serious snag [2]. There are lots of accidents that had occurred due to insufficient runway space. For an instance, Air India Express -811/812 landed 5200 ft from the beginning of Runway 24 of Mangalore International Airport in 2010. It overran, falling over a cliff and catching fire, leading to the demise of 158 lives [3]. Also, Cubana de Aviación Flight 1216 suffered an accident as a result of an overshoot on a wet runway and exiguous deceleration in 1999 [4]. All such accidents have occurred due to insufficient runway length. On the other hand, Air Traffic Control has become difficult to implement during rush hours. The number of commercial flights handled globally is 102,465 per day [5]. The traffic control has to be maintained at any cost, to avoid irregular handling of aircrafts. The aircrafts must wait till it is permitted to land during the time of which it loiters around the airspace, wasting a considerable volume of fuel. It is seen that the air traffic is continuously increasing since 2012 [6]. If this continues to exist, traffic management will be subjected to question in the future. A mutual solution to both these problems can be provided by reducing the landing distance.

2. Mechanics of Landing

In order to minimize the landing distance, it is essential to obtain an expression for the same. Figure 1 shows the forces associated with an airplane while landing, namely Lift (L), Drag (D), Thrust (T), Weight (W), Rolling Friction (R) [7].

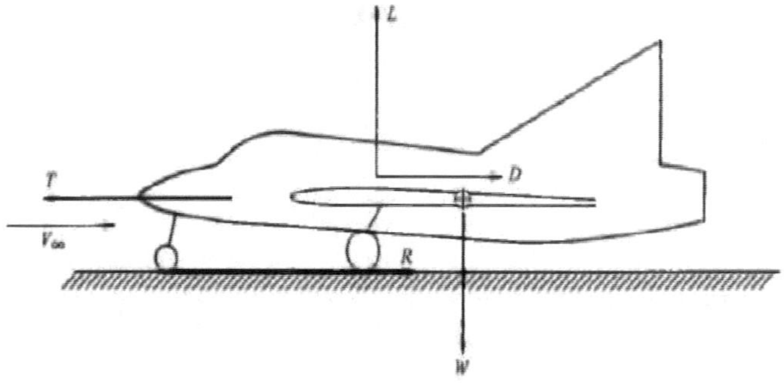

Figure 1. Forces on a flight while landing

The instantaneous acceleration acting in the opposite direction, is given by Newton's equations of motion [7],

$$F = T - \left(D + \mu_r(W - L)\right) \tag{1}$$

where μ_r is the *friction coefficient* whose value is 0.4 for a paved runway [7]. During landing, the thrust is theoretically zero [7]. Hence,

$$F = -\left(D + \mu_r(W - L)\right) \tag{2}$$

Substituting equation (2) in $s_L = -\frac{mV_T^2}{2F}$ [7], we have

$$s_L = \frac{1.69W^2}{\rho S g C_{L,max}\left(D+\mu_r(W-L)\right)_{0.7V_T}} \tag{3}$$

Newfangled aircrafts make use of thrust reversal (T_R) during the landing ground roll, by ducting air from the jet engines and blowing it in the upstream direction, opposite to the usual downstream when normal thrust is produced [7]. This aids the deceleration and shortens the ground roll. Hence equations (2) and (3) become,

$$F = -(T_R + D + \mu_r(W - L))_{0.7V_T} \tag{4}$$

and,

$$s_L = \frac{1.69W^2}{\rho S g C_{L,max}\left(T_R+D+\mu_r(W-L)\right)_{0.7V_T}} \tag{5}$$

respectively. The above expression for s_L is the minimum landing distance required by an ideal aircraft to complete its landing.

3. Decelerating Planar Field

It has been seen that, though the reverse thrusters have been deployed, overrun continues to occur. To decrease the landing distance, thereby reducing the air traffic and hence the accidents due to over-run, this paper proposes the solution in this section.

3.1. Generation of field

The deceleration concept is based on Lenz's law and production of Foucault currents (loops of electrical current induced within conductors by a changing magnetic field in the conductor in planes, perpendicular to the field [8]). Superconductors (red bars) are laid below the asphalt layer (green plane) as in figure 2. The landing aircraft (grey cylinder) moves along $\vec{V}$. The wings are neglected as they are parallel to the field.

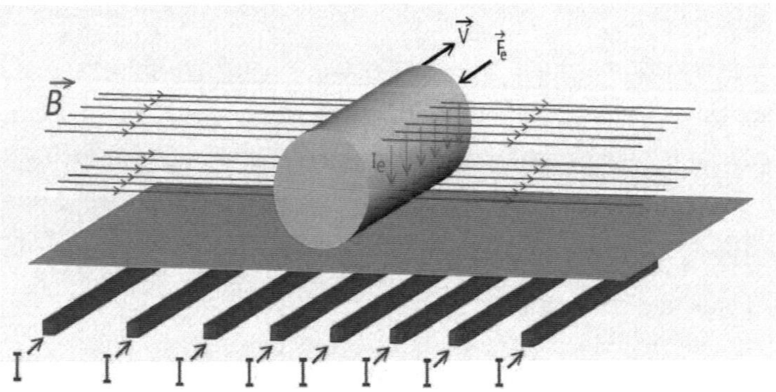

Figure 2. Flux, force and eddy current Representation

When current (I) is passed through the superconductors, the field ($\vec{B}$) is produced horizontally. The direction of $\vec{B}$ is given by cork rule. By Fleming's right hand rule, these Foucault currents are produced normally. Now, according to Fleming's left hand rule, a decelerating force ($\vec{F_e}$) is generated in the opposite direction to the aircraft's movement. Such a huge amount of field can be generated by means of high temperature superconductors [9]. The charges are restricted to the surface as the aircraft body is a Faraday's cage [10].

3.2. Expression for Decelerating force ($\vec{F_e}$)

The decelerating force aids the drag (D), adding up to the net opposing forces. This results in the minimization of the landing distance. If $e = BlV_i$ is the emf induced in a straight conductor [11], the differential form of Lorentz force is given by [12],

$$d\vec{F_e} = \frac{e}{\rho_e l} B \sin\theta \, d\theta \tag{6}$$

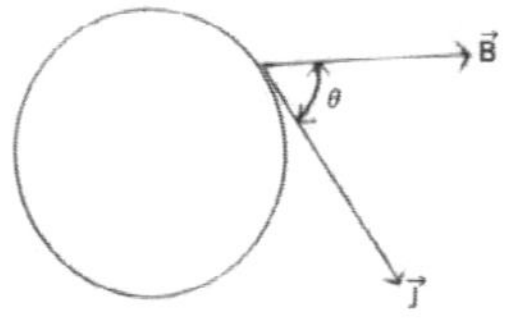

Figure 3. Eddy current density and flux density- Vector diagram

Figure 3 shows the cross sectional path of the cylinder showing the respective vectors. Considering only one section of the closed path of the reverse eddy current (i.e) current flows from $\theta = 0$ to $\theta = \pi$.

$$F_e = 4\frac{V_i B^2}{\rho_e} A \tag{7}$$

It is seen that the force generated is proportional to the square of the flux density (B).

3.3. Interface constraints

The superconductors produce the magnetic field (B), approximately given by [12],

$$B = \frac{\mu_g I}{2\pi r} \tag{8}$$

where μ_g is ground permeability. Figure 4 shows the vector component representation of flux density ($\overrightarrow{B}$) as the flux travels from ground to air. $\overrightarrow{B_g}$ is the flux density in the ground medium and $\overrightarrow{B_a}$ is that of the air medium. $\overrightarrow{B_g}$ and $\overrightarrow{B_a}$ make an angle $\emptyset$ and α with the barrier, respectively. The flux density varies normally and tangentially (i.e) $\overrightarrow{B_g}$ as $\overrightarrow{B_{gn}}$ and $\overrightarrow{B_{gt}}$ and $\overrightarrow{B_a}$ as $\overrightarrow{B_{an}}$ and $\overrightarrow{B_{at}}$. From figure 4, these are given as $|\overrightarrow{B_{gn}}| = |\overrightarrow{B_{an}}| = |\overrightarrow{B_g}| \sin \emptyset$, $|\overrightarrow{B_{gt}}| = |\overrightarrow{B_g}| \cos \emptyset$, $|\overrightarrow{B_{at}}| = \frac{\mu_0}{\mu_g} |\overrightarrow{B_{gt}}|$ and $|\overrightarrow{B_{at}}| = \frac{\mu_0}{\mu_g} |\overrightarrow{B_g}| \cos \emptyset$ and the resultant flux density is given by,

$$|\overrightarrow{B_a}| = \left[\left(\frac{\mu_0}{\mu_g}\right)^2 \cos^2 \emptyset + \sin^2 \emptyset\right]^{1/2} |\overrightarrow{B_g}| \tag{9}$$

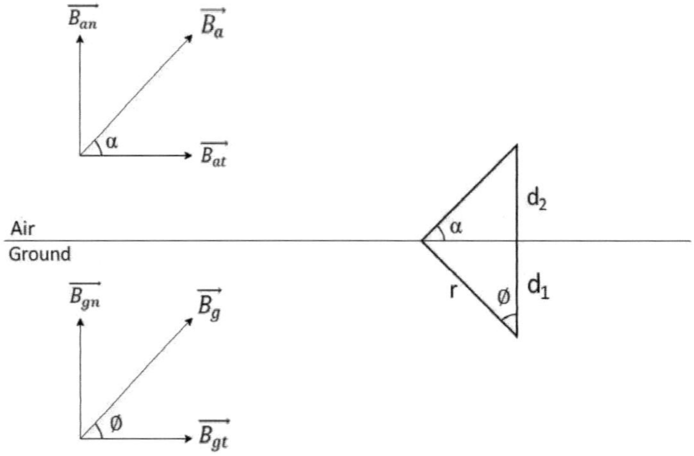

Figure 4. Vector representation of ground-air Barrier

From the boundary conditions [12], $\cot \alpha = \frac{\mu_0}{\mu_g} \cot \emptyset$. Substituting the known values,

$$r = \left[d_1\, d_2 \frac{\mu_0}{\mu_g} + {d_1}^2\right]^{1/2} \tag{10}$$

From ratios of trigonometry, $\cos \emptyset = \frac{d_1}{r}$; $\sin \emptyset = \frac{[r - {d_1}^2]^{1/2}}{r}$. Substituting the values of r, $cos\phi$ and $sin\phi$, $|\overrightarrow{B_a}|$can be expressed as,

$$|\overrightarrow{B_a}| = \frac{I\mu_g [\mu_0]^{1/2}}{2\pi} \frac{[\mu_0 {d_1}^2 + \mu_g d_1 d_2]^{1/2}}{\mu_g {d_1}^2 + \mu_0 d_1 d_2} \tag{11}$$

Substituting the value of $\overrightarrow{B_a}$ in equation (7), we obtain

$$F_e = 4\left(\frac{V_i l^2 \mu_g^2 \mu_0}{(2\pi)^2 \rho_e} \frac{\mu_0 {d_1}^2 + \mu_g d_1 d_2}{\left(\mu_g {d_1}^2 + \mu_0 d_1 d_2\right)^2}\right) A \qquad (12)$$

The g-force bearable by an average passenger inside an aircraft is 1.16 g [13]. But in case of fighter jets, the pilots alone are considered. They are trained to endure higher g-forces [14]. Now, equation (4) is updated as,

$$F = -(T_R + D + \mu_r(W - L) + F_e)_{0.7V_T} \qquad (13)$$

Now the minimum landing distance equation with respect to equation (13) is,

$$s_L = \frac{1.69W^2}{\rho S g C_{L,max}(T_R + D + \mu_r(W - L) + F_e)_{0.7V_T}} \qquad (14)$$

The above expression for s_L gives the minimum landing distance in terms of F_e. All the other quantities in equation (14) are fixed. By varying the value of F_e, s_L is varied inversely.

4. Results and discussions

Let us consider an Airbus A380 for analysis. Table 1 shows the values of the forces associated with landing and other data [15]. The aircraft body is usually made up of aluminium and other composites to bring about homogeneity [16].

Table 1. Specifications of an Airbus A380

6S	$845m^2$
T_R	160*kN*
W	3861*kN*
L	1947*kN*
D	31.13*kN*
V	68.056 *m/s*

Three cases are simulated namely the normal landing (red), planar field landing (blue) and the imprudent landing (green), as shown in figure 5. Now let us consider that an external force is generated with the help of planar field when an Airbus A380 is landing. For typical landing (red), the landing distance is 1931.8 *m*. The same A380 associated with planar fields (blue), takes only 1072.6 *m* (55.52% of typical distance). The corresponding value of F_e is calculated to be 266.9 kN. The blue curve in the velocity graph shows a linear decrease which indicates a smooth landing. Taking into account the passenger comfort, the maximum reducible distance (green) is found to be 500 m from simulations. But the g-force is just below the maximum bearable limit of 1.16 as shown in figure 6. This can be implemented under absolutely emergency conditions, but not recommended for normal cases. Due to this reason, it can be referred to as imprudent landing.

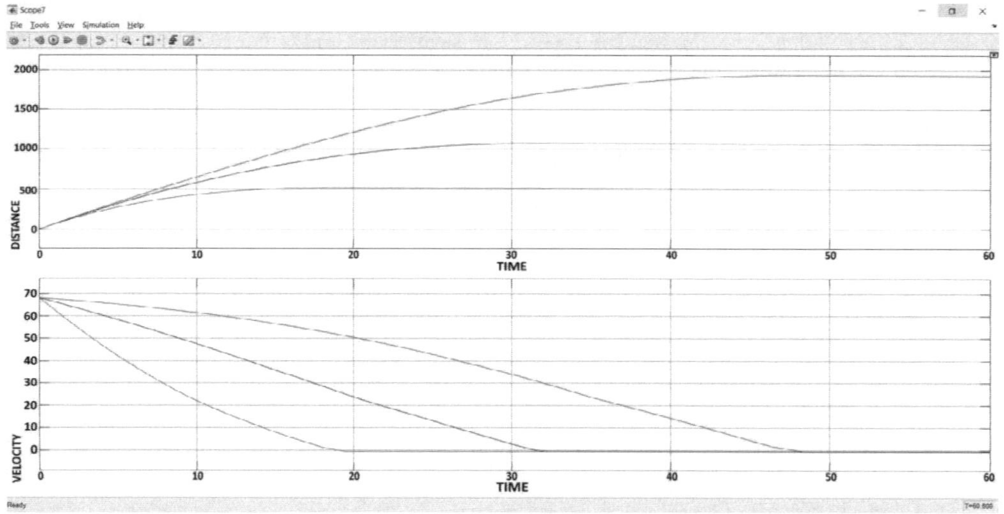

Figure 5. Distance (above) and velocity (below) variation while landing – for normal (red), planar (blue) and imprudent planar (green) landing of an Airbus A380

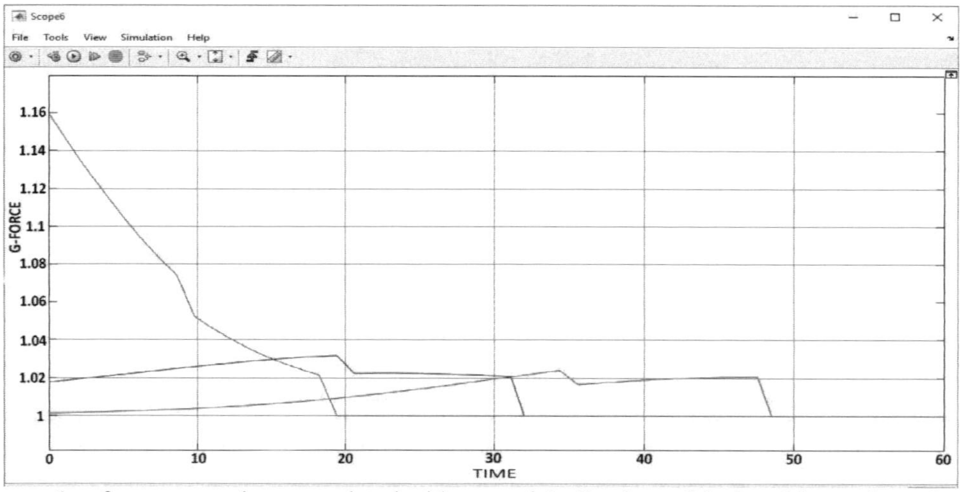

Figure 6. g-force comparison associated with normal (red), planar (blue) and imprudent planar (green) landing of Airbus A380

5. Conclusion

It has been verified that, on applying the concept of planar fields, it is possible to minimize the landing distance of an aircraft. Subsequently, overrun of aircrafts can be avoided and the accidents due to it can be circumvented, with the features of typical landing. As the landing will be completed quickly, air traffic becomes easy to manage. The volume of fuel that is wasted due to loitering, can be conserved. This idea can be applied to the aircraft carrier ships having shorter runways and the places minimum a distance is inevitable. The discussed braking technique is comparatively superior, as it is applied to the whole aircraft body and the stress is distributed uniformly. Hence, this technique is safe and electrically compatible as the field is restricted to the surface. The value of the externally generated force (F_e), is limited due to the g-force bearable by humans. However, this concept can be extended to the cases where F_e has no limit, which includes defense applications such as decelerating a high speed missile and hindrance to bombings.

Acknowledgements

We kindly express our gratitude to Mr. M Veerasundaram, Asst. Professor, Department of Electrical and Electronics Engineering, Sri Sai Ram Institute of Technology, Chennai, Mr. G Swaminathan, Asst. Professor, Department of Mechanical Engineering, Sri Sai Ram Engineering College, Chennai, Ms. D Ezhilmathi, Asst. Professor, Department of English, Sri Sai Ram Institute of Technology, Chennai and Dr. Tharinni Mai, Senior Librarian, Sri Sai Ram Institute of Technology, Chennai for their moral support and encouragement in completing and developing this paper. We also thank our parents for supporting us throughout this technical venture.

References

[1] Aviation Knowledge, *Runway overruns 2009* (Available: http://aviationknowledge.wikidot.com/aviation:runway-overruns; Accessed: Jun. 1, 2016)

[2] M Ayres Jr., H Shirazi, R Carvalho, J Hall, R Speir, E Arambula, R David, D Wong and J Gadzinski 2011 *Improved Models for Risk Assessment of Runway Safety Areas* ACRP Report 50 (Washington: Transportation Research Board)

[3] B N Gokhale *Accident to Air India Express Boeing 737-800 Aircraft VT-AXV on 22nd May 2010 at Mangalore* (New Delhi: Ministry of Civil Aviation)

[4] 2000 *Official accident investigation report on Cubana de Aviación Flight 1216, Aurora Airport, Runway 19, Guatemala City, Guatemala. Dec. 21, 1999* rep f-di991221e (Guatemala City: Civil Aviation)

[5] G Garfors 2014 *Globetrotting Galore by Gunnar Garfors* (Available: http://www.garfors.com/2014/06/100000-flights-day.html; Accessed: Jun. 13, 2016)

[6] The Flying Engineer 2016 *The Indian aviation 2015 growth story deciphered* (Available: http://theflyingengineer.com/views/the-indian-aviation-2015-growth-story-deciphered/; Accessed: Jun. 3, 2016)

[7] J D Anderson 2011 *Introduction to Flight* 7th ed (New York: McGraw Hill) pp 458-466

[8] J Bird 2013 *Electrical circuit theory and technology* 5th ed (London: Routledge) p 95

[9] A P Durajski, R Szczęśniak and L Pietronero 2015 *High-temperature study of superconducting hydrogen and deuterium sulfide* (Annalen der Physik vol 528 no. 5) ed G W Fuchs (Weinheim: Wiley-VCH) pp 358–364

[10] M Tooley and D Wyatt 2006 *Aircraft digital electronic and computer systems: Principles, operation and maintenance* ISBN 978-0-7506-8695-2 (Burlington: Butterworth-Heinemann) pp 344-345

[11] B D D Cullity and C D Graham 2007 *Introduction to magnetic materials* 2nd ed (Chichester: Wiley-Blackwell)

[12] M Sadiku 2009 *Principles of Electromagnetics* (New Delhi: OUP India) pp 5-12, 134-136, 242-249, 273-275, 295-297, 327-350

[13] J Wolkovitch, C W Lamont and D W Lochtie 1971 *Fundamental limitations on V/STOL terminal guidance due to aircraft characteristics* NASA CR-1901 (Los Angeles: Mechanics Research Inc.)

[14] S Slade 2009 *Feel the G's: The science of gravity and g-forces* (New York: Compass Point Books)

[15] Airbus S A S 2005 *Airbus A380 : Aircraft Characteristics, Airport and Maintenance Planning*

[16] J A DeRose, T Suter and T Hack, Eds 2012 *Aluminium alloy corrosion of aircraft structures: Modelling and simulation* (Southampton: Computational Mechanics)

IOP Conf. Series: Materials Science and Engineering **234** (2017) 012002 doi:10.1088/1757-899X/234/1/012002

Examining Mechanical Strength Characteristics of Selective Inhibition Sintered HDPE Specimens Using RSM and Desirability Approach

D. Rajamani[1] and Balasubramanian Esakki[2]

[1,2]Centre for Autonomous System Research (CASR), Department of Mechanical Engineering, Veltech Dr.RR & Dr.SR University, Chennai-600 062, India

[1]Corresponding author E-mail: rajamanitamil1991@gmail.com

Abstract. Selective inhibition sintering (SIS) is a powder based additive manufacturing (AM) technique to produce functional parts with an inexpensive system compared with other AM processes. Mechanical properties of SIS fabricated parts are of high dependence on various process parameters importantly layer thickness, heat energy, heater feedrate, and printer feedrate. In this paper, examining the influence of these process parameters on evaluating mechanical properties such as tensile and flexural strength using Response Surface Methodology (RSM) is carried out. The test specimens are fabricated using high density polyethylene (HDPE) and mathematical models are developed to correlate the control factors to the respective experimental design response. Further, optimal SIS process parameters are determined using desirability approach to enhance the mechanical properties of HDPE specimens. Optimization studies reveal that, combination of high heat energy, low layer thickness, medium heater feedrate and printer feedrate yielded superior mechanical strength characteristics.

1. Introduction

Selective Inhibition Sintering (SIS) is one of the novel and inexpensive AM techniques to fabricate parts on a layer-by-layer basis that has been introduced at University of Southern California, USA [1]. The crux of the SIS process is cohesion of polymer powders through sintering and inhibition at the part boundaries. SIS process has attractive features like rapid fabrication, use of wide range of processing materials (polymer, ceramic, and metal), enhanced sintering capability, and cost effective that leads to a great impact to RP community [2]. The following steps are involved in the production of SIS processed part:

1. Polymer powder is initially spreaded over the build platform through roller mechanism to build a first layer,
2. Inhibitor is deposited at the selected boundaries of built layer through print head nozzle, which is moving in X-Y plane,
3. Formulated layer is heated by an inexpensive heating system to perform sintering of polymer powder particles,

4. Layer by layer addition of materials with the above sequence will be repeated until the desired part profile is obtained. Finally, fabricated part is extracted from build chamber and post processing including removal of inhibitor, and cleaning of surplus materials from the part completes the SIS process.

Although SIS process is an economical and rapid way to fabricate functional parts, full-scale commercialization of this process has not gained much attention because of compatibility issues on the selection of process parameters. Asiabanpour *et. al.* [3] conducted few experiments in SIS process evaluation resulted that, layer thickness, heater energy and heater feed rate have affected the mechanical properties. In order to obtain maximum mechanical strength, it is essential to have complete control over the relevant process parameters [4]. Statistical design of experiments and empirical modelling of process parameters are efficient methodologies to improve the quality and performance characteristics [5-6]. Therefore, in this analysis an attempt has been made to model and optimize the SIS process parameters for maximizing mechanical strength of high density polyethylene (HDPE) specimens using response surface methodology and desirability approach.

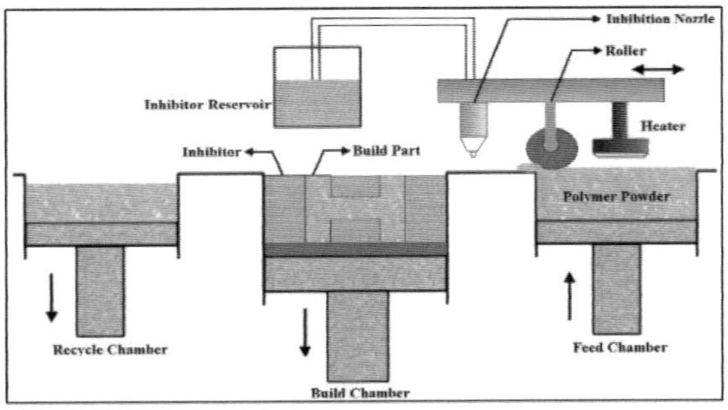

Figure1. Selective inhibition sintering process

2. Experimental methods

HDPE powders of particle size ranging from 25-40 μm is used as base powder and potassium iodide is considered as inhibitor throughout the study. Based on the existing literature [3], various SIS process parameters and their levels are accounted for part fabrication which is given in Table 1. The parts are built using a self-developed laboratory SIS machine. The schematic layout of SIS process is depicted in Fig.1. The working range of each process parameters is analyzed through inspecting the degradation of polymer and thermal stress distribution on specimen surface.

Table 1. SIS process control factors and their levels

Level	Layer thickness (mm)	Heater energy (J/mm^2)	Heater feedrate (mm/sec)	Printer feedrate (mm/min)
	A	B	C	D
Level 1	0.1	22.16	3	100
Level 2	0.15	25.32	3.25	110
Level 3	0.2	28.48	3.5	120

In order to plan and conduct the experiments, four-factor three-level response surface methodology (RSM) based box-behnken design [7] is considered and the corresponding design matrix is shown in Table 2. The tensile and flexural test specimens are fabricated with reference to ASTM standards D638 and D790, respectively. Structural strengths of SIS specimens are assessed under ambient conditions through Zwick/Roell automated material testing system with crosshead speeds of 2 mm/min. Experiments are performed and SIS part specimens mechanical strength characteristics are estimated. It is observed from Table 2 that, maximum tensile strength (TS) of 26.44 MPa and flexural strength (FS) of 64.88MPa is attained at a layer thickness of 0.1mm with heat energy of 28.48 J/mm^2, heater feed rate of 3.25mm/min and printer feed rate of 110mm/min.

Table 2. Experimental layout using box-behnken design to evaluate mechanical strengths

Run	Layer Thickness (mm)	Heater Energy (J/mm^2)	Heater Feed rate (mm/sec)	Printer Feed rate (mm/min)	Tensile Strength (MPa)	Flexural Strength (MPa)
1	0.20	25.32	3.50	110	23.05	40.46
2	0.15	28.48	3.00	110	21.29	56.64
3	0.10	25.32	3.50	110	22.53	36.34
4	0.15	25.32	3.25	110	25.55	55.63
5	0.20	25.32	3.25	100	24.39	46.35
6	0.10	25.32	3.25	120	24.94	47.18
7	0.15	25.32	3.50	120	21.53	47.2
8	0.15	25.32	3.00	120	20.41	43.11
9	0.15	28.48	3.50	110	25.92	53.57
10	0.15	25.32	3.25	110	25.12	54.55
11	0.15	28.48	3.25	100	24.64	53.13
12	0.15	22.16	3.25	100	23.97	48.9
13	0.20	25.32	3.00	110	20.86	44.18
14	0.15	22.16	3.25	120	21.29	45.04
15	0.15	22.16	3.50	110	19.34	39.24
16	0.20	22.16	3.25	110	23.77	52.57
17	0.15	25.32	3.25	110	25.29	55.22
18	0.10	25.32	3.00	110	25.87	57.6
19	0.15	25.32	3.50	100	21.62	45.4
20	0.15	25.32	3.25	110	24.9	52.22
21	0.15	25.32	3.25	110	24.7	52.1
22	0.15	25.32	3.00	100	24.97	53
23	0.15	28.48	3.25	120	25.25	54.19
24	0.10	25.32	3.25	100	24.85	52.11
25	0.10	28.48	3.25	110	26.44	64.88
26	0.10	22.16	3.25	110	23.8	48.01
27	0.20	28.48	3.25	110	24.97	48.94
28	0.15	22.16	3.00	110	24.47	47.71
29	0.20	25.32	3.25	120	22.9	35.77

3. Results and discussion

3.1. Statistical analysis and development of models

The statistical analysis and adequacy of developed models are tested using ANOVA (Analysis of Variance) at 95% significance level. The results of linear and quadratic order response surface model fitting for TS and FS in the form of ANOVA are depicted in Table 3. Statistical Design Expert v6.0 software is used to evaluate the effects of process parameters, data analysis, and quadratic model building. Further, the effectiveness of model is validated using co-efficient of determination (R^2) value. In the present work, R^2 value is 96.3% for TS and 86.7 % for FS is closer to 1, which has reasonable agreement and indicates adequacy of developed models. The adequate precision ratio for TS and FS is above 4 specifies adequate model discrimination.

Table 3. The ANOVA table for the fitted models

Source	Sum of square	d.f	Mean square	F	p	R^2	Adj.R^2	Pre.R^2	Adequate precision
TS									
Model	95.325	14	6.809	25.927	<0.0001	0.963	0.926	0.805	21.361
Total	99.002	28							
Residual	3.677	14	0.263						
Lack of fit	3.238	10	0.324	2.955	0.1539				
Pure error	0.438	4	0.109						
FS									
Model	1060.9	14	75.78	6.51	0.0006	0.867	0.734	0.271	9.937
Total	1223.9	28							
Residual	162.97	14	11.64						
Lack of fit	151.76	10	15.18	5.415	0.059				
Pure error	11.211	4	2.803						

The normal probability plot of the residuals of TS and FS are shown in Fig.2 (a-b). It reveals that the residuals are falling on the straight line that signifies the errors are distributed normally.

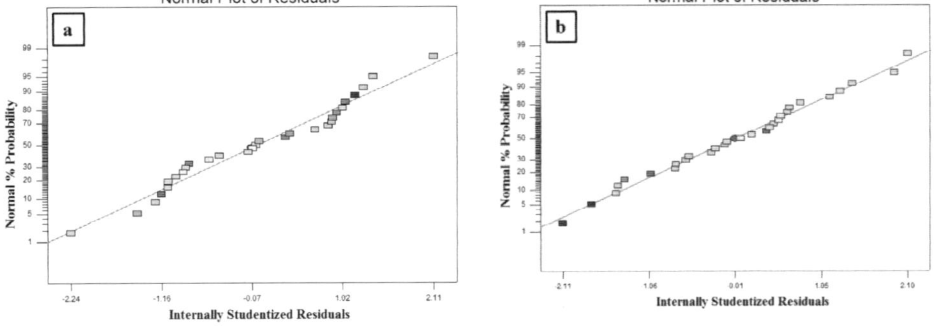

Figure2. Normal probability plot for TS and FS

From the above analysis after excluding the insignificant terms, the second-order quadratic models (i.e., in terms of actual values) for TS and FS are obtained and disclosed in Eqn.1 and 2.

$$\begin{aligned} Tensile\ strength &= 70.459 - 232.489 \times A - 10.352 \times B + 66.357 \times C - 0.032 \times D \\ &- 2.278 \times AB + 110.6 \times AC - 0.79 \times AD + 3.088 \times BC + 0.026 \times BD + 0.447 \times CD \\ &+ 11.6 \times A^2 - 0.037 \times B^2 - 32.556 \times C^2 - 0.009 \times D^2 \end{aligned} \quad (1)$$

$$\begin{aligned} Flexural\ strength &= -477.873 + 357.254 \times A - 9.911 \times B + 262.669 \times C + 4.1 \times D \\ &- 32.437 \times AB + 350.8 \times AC - 2.825 \times AD + 1.708 \times BC + 0.039 \times BD + 1.169 \times CD \\ &- 1427.97 \times A^2 + 0.123 \times B^2 - 76.999 \times C^2 - 0.039 \times D^2 \end{aligned} \quad (2)$$

Where A, B, C and D are layer thickness, heater energy, heater feedrate, and printer feedrate, respectively. Overall, statistical analysis reflected that the experimental values are fitted well within predicted ones and the accuracy of the model is adequate to further explore optimization studies on determining SIS process parameters.

3.2. Effect of process parameters on Tensile strength

The effect of layer thickness and heater energy on tensile strength is shown in Fig. 3a. It can be seen from the plot that tensile strength is maximum at higher heater energy, whereas it slightly decreases with increase in layer thickness. It is consistent with the fact that increase in heat energy improves the degree of sintering and flow of polymeric particles and a well-defined structure is formed. Hence, tensile strength of the specimen increased with an increase in heater energy. At a low level of layer thickness, the sufficient heat energy can be transferred to the powder particles resulted in better fusion of HDPE particles and achieved a more compact structure. When layer thickness increases, distortion effect dominates the bonding effect and decreases tensile strength of specimen. The interaction effect of heater feedrate and layer thickness on tensile strength is depicted in Fig.3b. It is observed that with an increase in layer thickness causes decrease in tensile strength from 25.39 to 22.15 MPa.

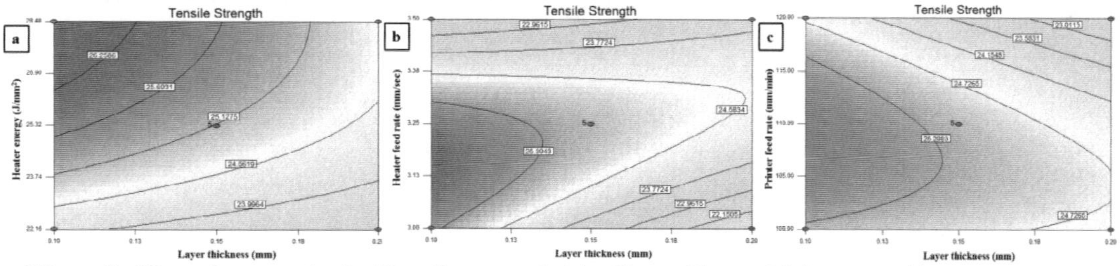

Figure3. 2D surface graphs for Tensile strength (a) effect of layer thickness and heater energy, (b) effect of layer thickness and heater feedrate, and (c) effect of layer thickness and printer feedrate

However, increase in heater feedrate improves the TS up to 3.28 mm/sec and then decreases. It can be justified with the fact that at maximum heater feedrate, the energy absorbed by powder particles is not uniform due to shorter heat time which causes improper sintering. Therefore, poor sintering decreases the tensile strength of sintered specimens. Fig. 3c demonstrates the effect of printer feedrate and layer thickness on tensile strength. It is inferred from simulation analysis that, tensile strength increases up to maximum value of 25.29 MPa and then decreases with increase in printer feedrate from 100 to 120 mm/min.

3.3. Effect of process parameters on Flexural strength

The response surface plots in Fig.4 (a-c) describe the effect of input process parameters on flexural strength. It can be understood that the effects of process parameters on evaluating flexural strength have similar trend as in the case of tensile strength of the sintered specimens. Fig. 4a reflects the effect of heater energy and layer thickness on flexural strength. It is observed that, flexural strength increases from 48.574 MPa to 60.94 MPa with increase in heater energy from 22.16 to 28.48 J/mm^2, whereas slightly decreases with the increase in layer thickness ranging from 0.1 to 0.2 mm. Similarly, the effect of heater feedrate and layer thickness on flexural strength has shown in Fig.4b. It can be seen that, flexural strength slightly increases with increase in heater feedrate from 3 to 3.25 mm/sec and layer thickness up to 0.14 mm and then decreases with the increases in these two parameters. Finally, the effect of printer feedrate and layer thickness depicted in Fig.4c indicates that, at low layer thickness of 0.1 mm and middle range of printer feedrate (110 mm/min) attained maximum flexural strength.

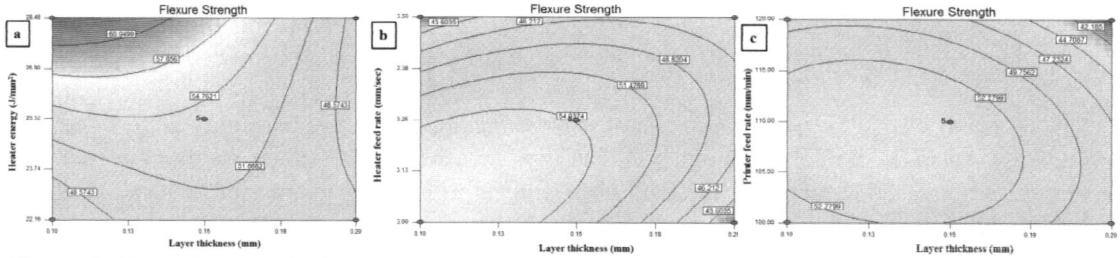

Figure4. 2D surface graphs for Flexural strength (a) effect of layer thickness and heater energy, (b) effect of layer thickness and heater feedrate, and (c) effect of layer thickness and printer feedrate

3.4. Optimization of SIS process parameters

Due to the presence of large number of input parameters in SIS and intricate stochastic process mechanism has tremendous effect in selection of optimal SIS process parameters to achieve improved mechanical properties. The present study emphasise a multi-objective problem to simultaneously optimize the process variables using desirability approach which is proposed by Derringer and Suich [8]. It is a unique optimization technique used for optimizing multiple quality characteristic problems in industries [9]. In a multi response domain, a measure of how the solution has satisfied the combined goals for all responses must be assured [10]. In this work, based on composite desirability optimization technique, the two responses i.e., tensile and flexural strengths have been optimized simultaneously using developed mathematical models as described in Eqn.1 and 2. The ramp desirability graph shown in Fig.5 describes the optimum points obtained through numerical optimization procedure. It reveals that, earlier predicted maximum tensile strength of 26.44 MPa and flexural strength of 64.33 MPa can be attained under the SIS process conditions such as layer thickness of 0.11 mm, heater energy of 28.48 J/mm^2, heater feedrate of 3.21 mm/sec, and printer feedrate of 112.53 mm/min.

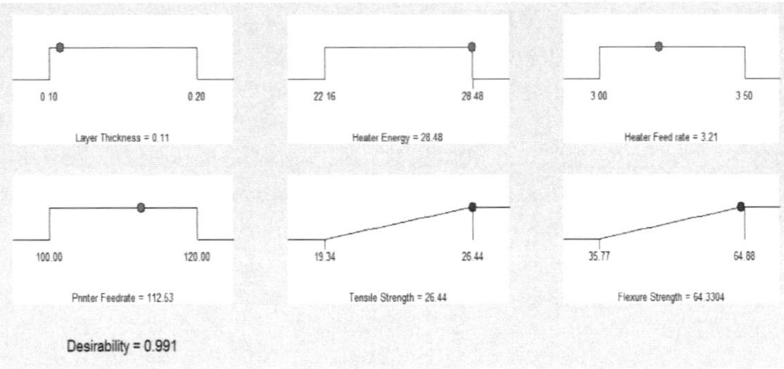

Figure5. Desirability ramp function for numerical optimization

4. Conclusions

HDPE specimens have been fabricated through SIS process. The mechanical properties including tensile strength and flexural strength are investigated using RSM with box-behnken design. Empirical models are developed to predict the mechanical properties of the sintered specimens. The following inferences are deduced from this study.

- Mathematical models of TS and FS are formulated to correlate the dominating input parameters, including layer thickness, heater energy, heater feedrate, and printer feedrate.
- The maximum tensile strength of 26.44 MPa and flexural strength of 64.33 MPa could be attained through considering optimal SIS process parameters layer thickness of 0.11 mm, heater energy of 28.48 J/mm^2, heater feedrate of 3.21 mm/sec, and printer feedrate of 112.53 mm/min.
- Influence of heater energy followed by layer thickness and printer feedrate has predominant effect on mechanical strength. However, heater feedrate has insignificant effect on parts strength.
- The higher heater energy of 28.48 J/mm^2 offered higher mechanical strength. On contrary, TS and FS decreased with increase in layer thickness from 0.1 to 0.2 mm.

Acknowledgement

The authors are grateful to the financial support from the Armament Research Board (ARMREB), Defence Research and Development Organization (DRDO), Government of India (Grant No. ARMREB/MAA/2015/167) to carry out this research work. The authors also thankful to The Association of Commonwealth Universities for providing financial support through Early Career Academic Grants.

References

[1] Khoshnevis B, Asiabanpour B, Mehdi Mojdeh, Kurt Palmer (2003) *SIS – a new SFF method based on powder sintering.* Rapid Prototyp. J. 9(1):30-36

[2] Khoshnevis B, Yoozbashizadeh M, Chen Y (2012) *Metallic part fabrication using selective inhibition sintering (SIS)*. Rapid Prototyp. J. 18(2):144-153
[3] Asiabanpour B, Khoshnevis B, Palmer K (2006) *Advancements in the selective inhibition sintering process development*. Virtual Phys. Prototyp. 1(1):43-52
[4] Selvamani S.T, Palanikumar K (2014) *Optimizing the friction welding parameters to attain maximum tensile strength in AISI 1035 grade carbon steel rods*. Measurement, 53: 10-21
[5] Negi S, Dhiman S, Sharma R.K (2014) *Investigating the surface roughness of SLS fabricated glass-filled polyamide parts using response surface methodology*. Arab J Sci. Eng. 39:9161-9179
[6] Bahram Asiabanpour, Kurt Palmer, Behrokh Khoshnevis (2004) *An experimental study of surface quality and dimensional accuracy for selective inhibition of sintering*. Rapid Prototyp J 10(3):181-192
[7] Tamilarasan A, Rajamani D (2017) *Multi-response optimization of Nd; YAG laser cutting parameters of Ti-6Al-4V superalloy sheet*. J. Mach. Sci. Technol. 31(2): 813 – 821
[8] Derringer G, Suich R (1980) *Simultaneous optimization of several response variables*. J Qual. Technol 12:214-219
[9] Gopalakannan S, Senthilvelan T (2013) *EDM of cast Al/SiC metal matrix nanocomposites by applying response surface method*. Int J Adv Manuf Technol 67:485-493
[10] Vijayan D, Rao V.S (2014) *A multi response optimization of tool pin profile on the tensile behaviour of age-hardenable aluminium alloys during friction stir welding*. Res. J. Applied Sci. Eng. Technol. 7(21): 4503-4518

IOP Conf. Series: Materials Science and Engineering **234** (2017) 012003 doi:10.1088/1757-899X/234/1/012003

Computational Study of Convective Heat Transfer in Circular Pipe using Trapezoidal Ribs

V K Patel[1], P Sharma[1] and K Saxena[1]

Applied Mechanics Department, MNNIT, Allahabad, UP 211004, India

vivek@mnnit.ac.in; pras21sharma@gmail.com; saxenakushal05@gmail.com

Abstract. Perturbations or interruptions provided in the passage of heat exchanger generate the vortices downstream. The formation of these natural vortices, augment local heat transfer abruptly. The effect on convective heat transfer enhancement and friction characteristics by providing trapezoidal ribs inside a circular pipe is computationally investigated in detail. Different variations of height, width and pitch of the ribs are used to optimize the rate of heat transfer through the pipe. Liquid water is employed as the working fluid. Input parameters of Reynolds Number ranging from 5000-60000 with axial flow along the pipe and constant heat flux of 50 W/cm^2 to the pipe surface is used. After validation with the existing literature, Realizable k-ε turbulent model with enhanced wall function is used in commercial CFD software ANSYS FLUENT. The outcome of the investigation shows that the ribs provided on the inside of the pipe surface enhance the turbulence in the flow and produce recirculation which disturb the thermal boundary layer behind the ribs and thus help in enhancing the rate of heat transfer through the pipe.

1. Introduction

Need of heat transfer enhancement in process industries, chemical industries, evaporators, power plants condensers, air-conditioners and refrigerators, radiators, electrical panels, etc. has always drawn the attention of the researchers. Augmentation of heat transfer through a heat exchanger can broadly be done by three different methods, viz. active, passive and compound method. In active method some external power input is given to enhance the heat transfer rate. In passive method no external power input is required to create turbulence or swirling of the fluid. Instead inserts are provided in the flow tubes or modification in the geometry of the surface of the fins or flow tubes is done to enhance the rate of heat transfer, but this brings pressure loss in the flow field as a drawback [1].

Mohammad et al. [2] in their computational study of transversely corrugated circular tubes with different geometrical parameters observed that Nusselt number (*Nu*) increases with increasing Reynolds number (*Re*) and decreasing pitch of the corrugation. On the other hand the, friction factor do not change with alteration in Reynolds number in corrugated tubes but friction coefficient reduces with increase in Reynolds number. A computational comparative study between Reynolds stress transport (RST) and direct numerical simulation (DNS) was put forwarded by Han et al. [3] on an outward convex corrugated steel tube with helium as the fluid. They proposed that RST model matches closely with DNS model numerical results and hence is more suitable than any other Reynolds-Averaged Navier Stokes model for the study of flow and heat transfer characteristics of corrugated tubes. Kathait and Patil [4] experimentally studied the thermo hydraulic performance of a rectangular discrete ribbed copper tube with water as the working fluid. The findings show that the values of Nusselt number and friction factor increase by increasing the pitch to rib height ratio. The maximum enhanced value of 2.73 for Nusselt number at pitch to rib height ratio of 10 was reported. Liu et al. [5] computationally investigated the heat transfer and flow characteristics of shell side flow of a rod baffle heat exchanger with single, double, triple and quadruple start spirally corrugated tube with water as the fluid. The Nusselt number enhanced to 1.2 times for single start spiral tube at 18000 Reynolds number. The successive increase in number of start of spiral corrugation increased the corresponding heat transfer slightly. A smooth two start spirally corrugated tube with water as the flowing fluid, was analyzed computationally at low Reynolds number from 100-1300 by Kareem et al. [6]. The effect of ratio of corrugation height to tube diameter and corrugation pitch to tube diameter were studied. At Reynolds number greater than 700, increase in friction factor was found to be much greater than the increase in heat transfer. An augmentation in heat transfer of 19%-71% was reported with a particular severity of corrugation.

The literature review presented here suggests that the emphasis is given to the flow with low Reynolds number. The present study employs Reynolds number in the range of 5000-60000 and water as the fluid. Trapezoidal ribs protruded on the inner surface of the pipe are used as the turbulators to enhance the turbulence in the flow. The reverse flow developed by ribs can enhance the heat transfer coefficient and the momentum transfer and has a consequence in the form of higher pressure drop along the flow. The enhancement relies on the strength of the recirculation and the prolongation of the reattachment.

2. Methodology

2.1. Geometrical Model Description

The analysis is performed in 2-D, considering the flow to be axisymmetric about the axis of the pipe. The trapezoidal ribs were protruded circumferentially along the length of the pipe at different pitches for different cases. Figure 1(a) shows the geometry of rib and figure 1(b) shows the schematic of the flow with different boundaries. The diameter of the pipe is 10 mm; entrance length is taken as 200 mm considering the fully developed flow at the test section. Length of the test section is 100 mm while the exit section length is 50 mm.

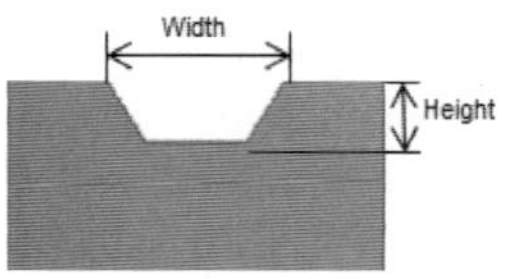

(a)

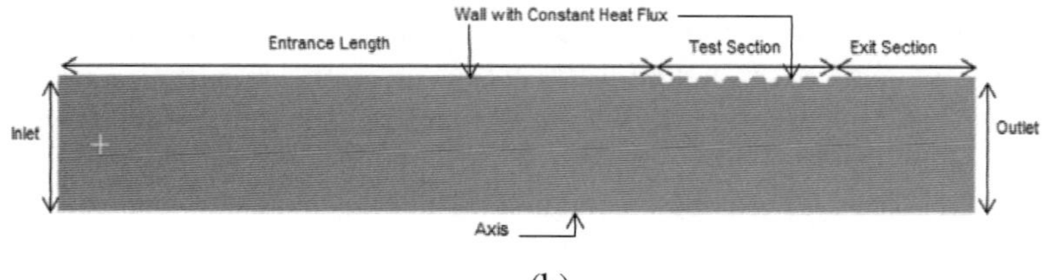

(b)

Figure 1. Trapezoidal rib along the fluid flow.

2.2. Discretization

A structured grid was adopted for the analysis. As wall gradient has a significant role in heat transfer problems a finer mesh is adopted by proper biasing factor near walls to capture the properties of flow in viscous sub layer. The nearest wall distance governed by y^+ approach was considered which lies below 5. A grid independent test was performed to verify the results. Grid test was performed with width of the ribs 0.5mm, height of the ribs 0.25mm and pitch of the ribs 5mm. Grid sizes employed was 64929, 172959 and 458976. Pressure and Velocity distribution was considered for grid independency test. Grid with 458976 numbers of elements was adopted for the simulation after the grid independent test as the y+ value lied below 1 for this.

2.3. Boundary Conditions and Material Selection

Inlet was given selected as velocity inlet with different velocities to account for the desired Reynolds number. For turbulent specification hydraulic diameter of 10mm and the turbulent intensity of 3% were selected. At outlet zero gauge pressure outlet was selected. The upper edge was selected as wall with no slip condition and the test section length was given a constant heat flux of 50 W/cm^2. With centerline as axis the axisymmetric solver was selected. Properties of water are shown in table 1.

Table 1. Properties of the liquid water

Properties	Specifications
Density (kg/m3)	998.2
Specific Heat (J/kg-K)	4182
Thermal Conductivity (W/m-K)	0.6
Viscosity (kg/m-s)	0.001003

2.4. Solver Settings

For pressure velocity coupling SIMPLE scheme was used. Pressure equation has been discretized using Standard. Momentum, Turbulent Dissipation Rate, Turbulent Kinetic Energy and Energy equations has been discretized using second order upwind scheme.

2.5. Governing Equations

Following are some important equations for the fluid flow in a pipe with heat exchange in focus [7].

Reynolds number $Re = \frac{\rho V_{in} D_h}{\mu}$,

Nusselt number $Nu = \frac{h D_h}{k}$,

Friction factor $f = \frac{\Delta p}{\frac{\rho V_{in}^2}{2}\frac{A_T}{A_{min}}}$

Newton's law of cooling: $q_{conv} = h(T_s - T_\infty)$

Continuity equation in two dimensions: $\frac{\partial u}{\partial x} + \frac{\partial v}{\partial y} = 0$

Navier Stokes Momentum equation:

X-momentum: $u\frac{\partial u}{\partial x} + v\frac{\partial u}{\partial y} + \frac{1}{\rho}\frac{\partial p}{\partial x} = \vartheta\left(\frac{\partial^2 u}{\partial x^2} + \frac{\partial^2 u}{\partial y^2}\right)$

Y-momentum: $u\frac{\partial v}{\partial x} + v\frac{\partial v}{\partial y} - \frac{1}{\rho}\frac{\partial p}{\partial y} = \vartheta\left(\frac{\partial^2 u}{\partial x^2} + \frac{\partial^2 u}{\partial y^2}\right)$

Transport equation for k: $\frac{\partial(\rho k u_j)}{\partial x_j} = \frac{\partial}{\partial x_j}\left[\left(\mu + \frac{\mu_t}{\sigma_k}\right)\frac{\partial k}{\partial x_j}\right] + G_k - \rho\varepsilon$

Transport equation for ε: $\frac{\partial(\rho\varepsilon u_j)}{\partial x_j} = \frac{\partial}{\partial x_j}\left[\left(\mu + \frac{\mu_t}{\sigma_\varepsilon}\right)\frac{\partial \varepsilon}{\partial x_j}\right] + \rho C_1 S_\varepsilon - \rho C_2 \frac{\varepsilon^2}{k + \sqrt{v\varepsilon}}$

Energy Equation $u\frac{\partial T}{\partial x} + v\frac{\partial T}{\partial y} = \alpha\left(\frac{\partial^2 T}{\partial x^2} + \frac{\partial^2 T}{\partial y^2}\right)$

2.6. Validation

Some of the cases of variation in dimensions of the ribs as discussed by Mohammed et al. [6] were adopted for comparing their average Nusselt number with the present simulated results. The percentage error obtained in the different cases gives the clarification of the validated results with the Mohammed et al. [6]. For validation rib with width 0.5mm, pitch 5mm and height 0.25mm similar to the Mohammed et al. [6], was generated. Figure 2 shows a minor deviation in results of Average Nusselt number to Reynolds number reported from the referenced work, owing to the difference in the grid size, y$^+$ value and the turbulence model selected.

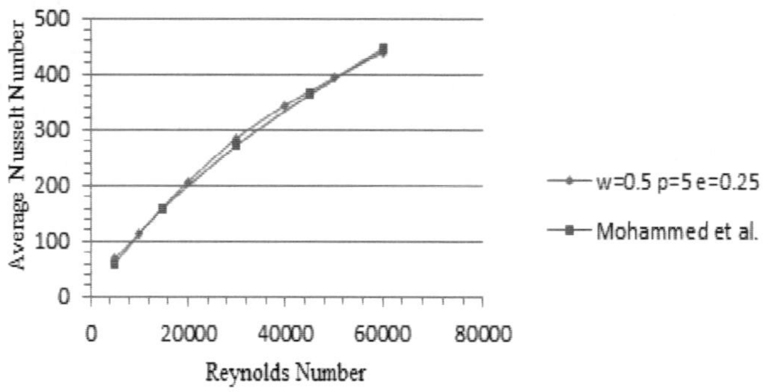

Figure 2. Validation of the CFD code.

3. Results and Discussion

Table 2 shows the different combinations that have been used to carry out the parametric investigation. Trapezoidal ribs were tested for five variations.

Table 2. Parametric description of the profiles (Dimensions in mm).

Case	Width	Height	Pitch
A	0.5	0.25	5
B	0.5	0.25	10
C	0.5	0.25	15
D	0.5	0.75	15
E	0.5	1	15

Figure 3 and 4 show the effects on average Nusselt number and the friction factor against variation in Reynolds number by varying the pitch of the ribs (5,10,15 mm), keeping width (0.5 mm) and height (0.25 mm) constant. It can be seen at higher Reynolds number the incoming velocity increases which disturbs the flow resulting in better heat transfer. With an increase in the periodicity of the ribs along the flow in pipe, the thermal boundary layer tends to get thinner and thinner as compared to flow in a plain pipe. This thinning of thermal boundary layer helps greatly in enhancing the rate of heat transfer. It is evident from figure 3 that Nusselt number increases in Nusselt number with a decrease in the pitch of the ribs and increasing the Reynolds number. On the contrary, friction factor reduced with the rise in Reynolds number for all the cases. By decreasing the pitch of the ribs the roughness of the flow field increased and consequently pressure drop and friction factor increased quite predictably.

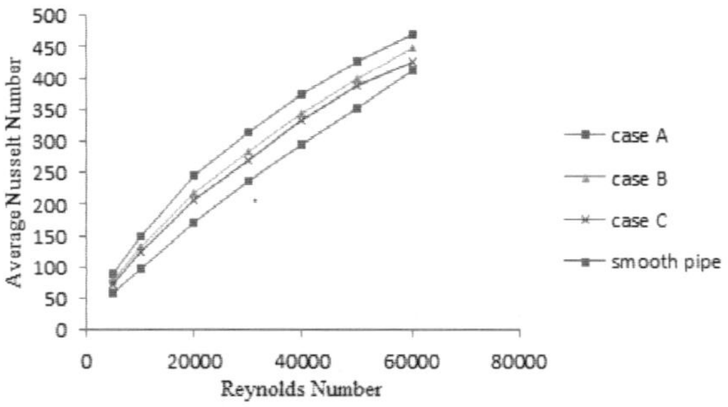

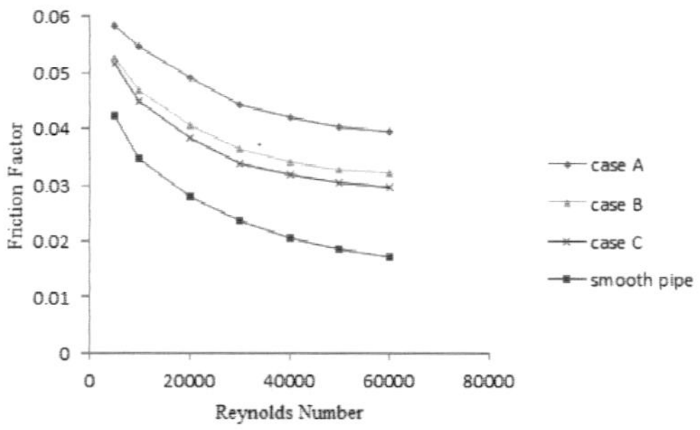

Figure 3. Variation of Nu with Re. **Figure 4.** Variation of f with Re.

Figure 5 and 6 show the effect on average Nusselt number and friction factor at different Reynolds number by varying rib height (0.25, 0.75, 1 mm) keeping width (0.5 mm) and pitch (15 mm) constant.

IOP Conf. Series: Materials Science and Engineering **234** (2017) 012003 doi:10.1088/1757-899X/234/1/012003

It is evident that with an increase in Reynolds number the average Nusselt number also increased. Similar trend for Nusselt number was seen by increasing the height as were reported by decreasing the pitch of the ribs. Furthermore, increasing the height of the ribs disturbs the whole flow field which causes enhanced heat transfer at the cost of enhanced drop in pressure. As shown in figure 6, the friction factor is increasing with an increase in the height of the ribs. Case E being the most prominent case as the height of the rib is highest in it.

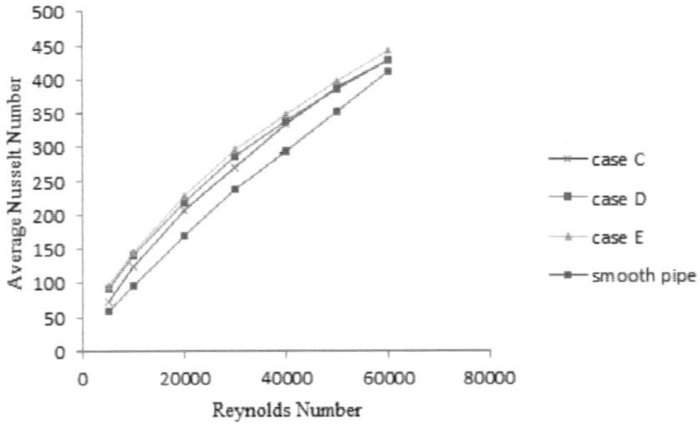

Figure 5. Variation of Nu with Re.

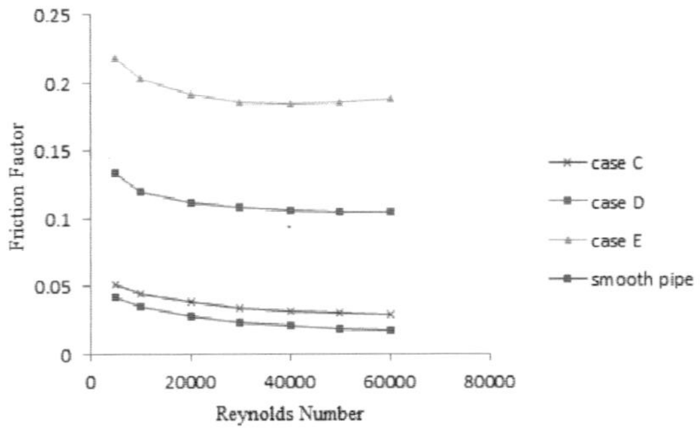

Figure 6. Variation of *f* with Re.

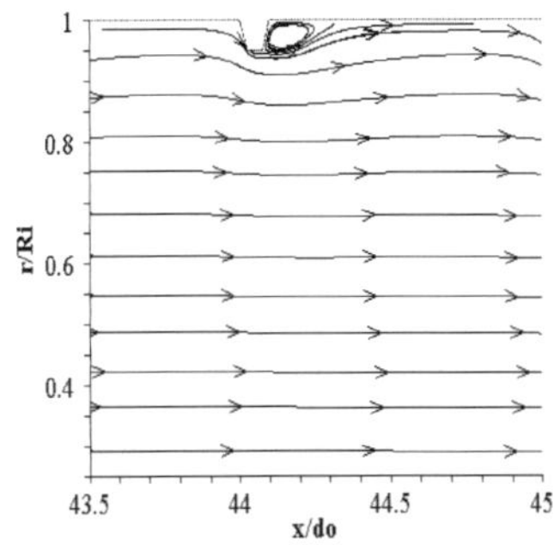

(a) Case A

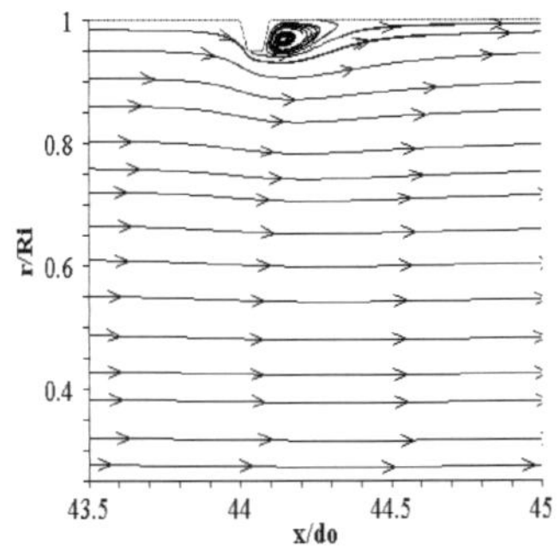

(b) Case B

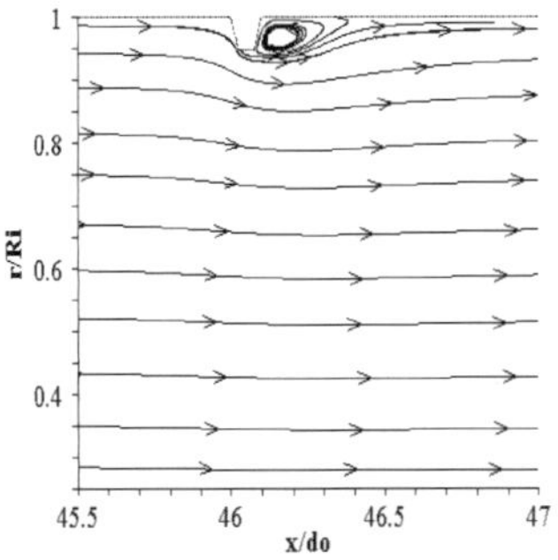

(c) Case

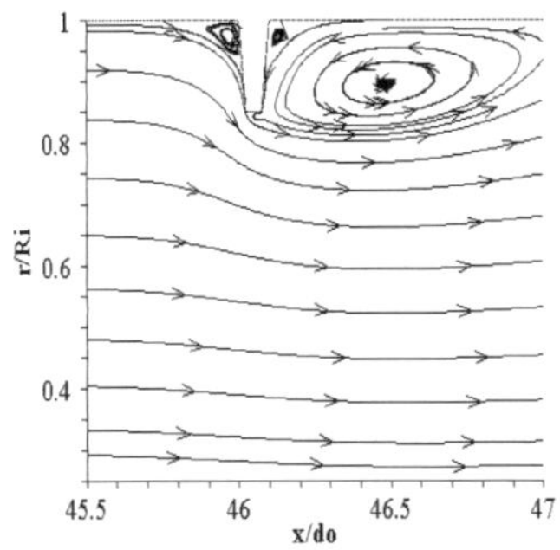

(d) Case D

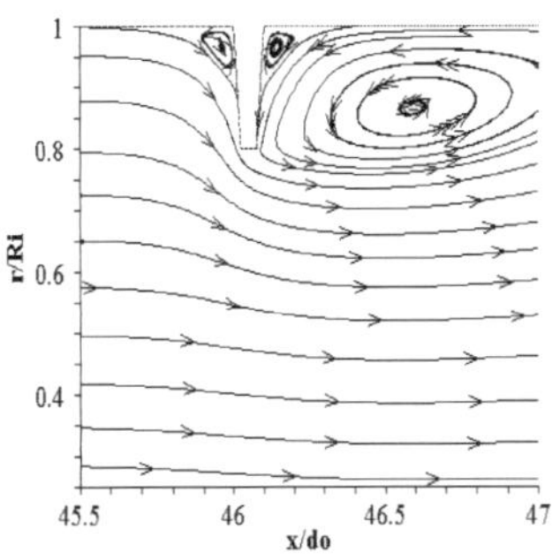

(e) Case E

Figure 7. Streamlines plot behind ribs.

With the change in Reynolds number, change in friction factor is seen to be little compared to variation in Nusselt number. Figure 6 shows the height of the ribs is a prominent factor considering the magnitude of the friction factor. Figure 7 illustrates the streamlines of the flow field. With an increase in height of the ribs more pronounced recirculation is taking place behind the ribs as shown in figure 7 (d) and 7 (e). Figure 8 shows the axial velocity distribution contours which demonstrated the presence of negative axial velocity which marked the presence of recirculation zone behind the ribs in the flow. Consequently, it can be clearly observed that the ribs having larger height shows bigger wake region behind ribs.

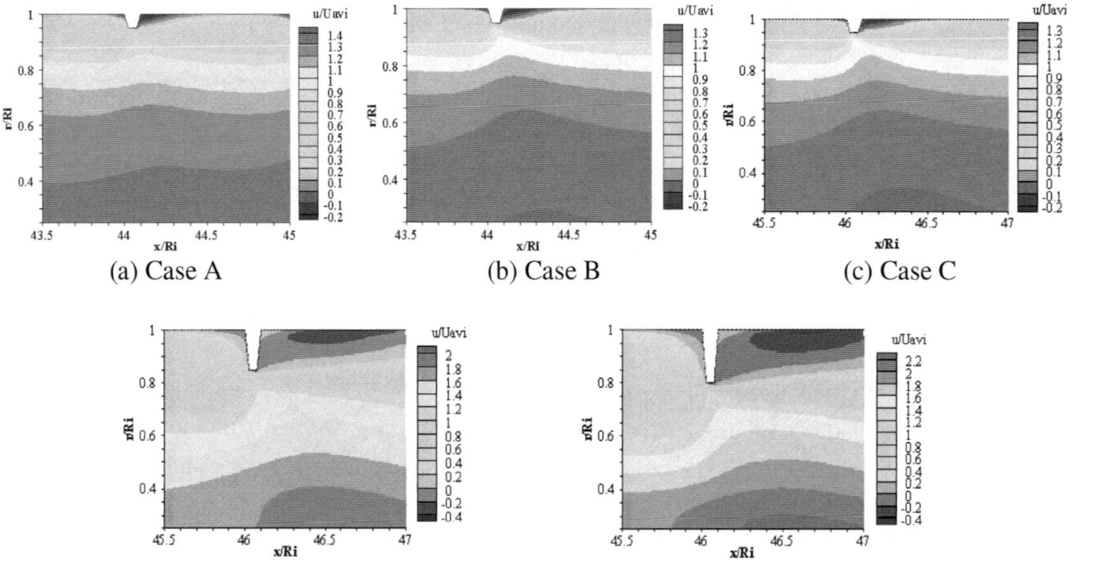

(a) Case A (b) Case B (c) Case C

(d) Case D (e) Case E

Figure 8. Axial velocity distribution.

The overall observation of the variations of pitch and height in the geometry of the ribs confirm that the change in pitch gives better results than changing the height of the ribs as far as rate of heat transfer and the friction factor are concerned. For the same increase in Nusselt number at a given Reynolds number, the friction factor increases by as much as 400% by increasing the height compared to decreasing the pitch of the ribs. However a quantitative analysis shows that the case which is showing maximum heat transfer enhancement (*i.e.* case C) shows 56% increase in Nusselt number at 10000 Reynolds number.

4. Conclusion

The immergence of a recirculation zone just behind the ribs disturbs the thermal boundary layer which enhances the rate of heat transfer. Nusselt number enhances with reduction in the pitch and by increasing the height and width of the ribs, resulting augmentations in convective heat transfer. Friction factor decreases with increase in pitch and reducing the height and width of the ribs. Friction factor varies very little as compared to Nusselt number by increasing Reynolds number.

References

[1] Wcbb R L 1994 *Principles of Enhanced Heat Transfer* (New York-Wiley).
[2] Mohammed H A, Abdalrazzaq K A and Sheriff J M 2013 *Int. Comm. Heat Mass Transfer* **44,** 116-26.
[3] Han H Z, Li B X , Li F C and He Y R 2013 *Computers & Fluids* **91,** 107-29.
[4] Kathait P S and Patil K 2014 *Applied Thermal Engineering* **66,** 162-70.
[5] Liu J J, Liu Z C and Liu W 2014 *International Journal of Thermal Sciences* **89,** 34-42.
[6] Kareem Z S, Jaafar M N M, Lazim T M, Abdullah S and Wahid A F A 2015 *Alexandria Engineering Journal* **54,** 415-22.
[7] Incropera F P, DeWitt D P, Bergman T L and Lavine A S **6th** Ed. *Introduction to heat transfer* (New Jersey-John Wiley & Sons, Inc).

IOP Conf. Series: Materials Science and Engineering **234** (2017) 012004 doi:10.1088/1757-899X/234/1/012004

Development of Gas Dynamic Probe for High Total Temperature Measurement in High Speed Flow

Tahzeeb Hassan Danish[1], Yash Mistry[2], Sathiyamoorthy K[1], Srinivas J[1] and Pratheesh Kumar P[1]

[1]Propulsion Division, NWTC, CSIR-National Aerospace Laboratories, Bangalore, Karnataka, India
[2]Graduate, IIAEM, Jain University, Bangalore, Karnataka, India

thdanish@nal.res.in

Abstract. Total temperature measurement in high speed and high temperature exhaust of air breathing engines like scramjet, ramjet and turbojets with afterburner is a challenging problem. The temperature of such hot gases is approximately 2000-2200 K. Conventional technique of temperature measurement by thermocouple suffers oxidation, doesn't withstand aerodynamic load, and lacks robustness. Unconventional technique like optical method is expensive. These limitations have withheld the temperature measurement in high speed high temperature exhaust streams. Present work describes the design and development of a water cooled gas dynamic probe for total temperature measurement of high speed and high temperature gases. The probe consists of two choked nozzles in series. The hot flow is cooled after passing through the first nozzle, in a small settling chamber, using a heat exchanger. The flow from settling chamber is accelerated to Mach 1 in second nozzle. Using gas dynamic relations and measured parameters, the incoming flow total temperature is obtained. The probe has been calibrated for probe constant at total temperature range of 1200 to 1700 K at inlet Mach number 2. Probe constant varies proportional to the square of incoming flow total temperature. First restriction with C-D nozzle gave higher probe constant value than convergent nozzle.

1. Introduction

Supersonic Combustion Ramjet engines (Scramjets) are the futuristic and most promising candidates in air breathing propulsion systems. Scramjets are similar to ramjets in operation but the combustion takes place at supersonic speeds. A scramjet vehicle operates efficiently at Mach number above 5. The operation of Scramjet engine is predominately based on the production and utilization of thermal energy due to combustion. Precise and reliable instruments for stagnation temperature and pressure measurements in the most hostile sections of the scramjet combustor are needed to improve engine designs, for the active control of the engine and for engine health monitoring. The enormous rise in temperature and high velocity at the exit of the scramjet combustor makes the temperature measurement very challenging task. Thermometry is fundamental to the control and monitoring of this process and hence the correct operation of engine. A rise in temperature often results in malfunctioning of equipment. Accurate and robust temperature measurement and control are therefore critical for maintaining and optimizing engine performance. Lack of information about the scramjet combustor exit temperatures affects the accuracy of predictions of engine performance, efficiency, component life and the development and maintenance costs.

Different types of instruments and sensors are used for temperature measurements in scramjets. These include thermocouples, infrared thermometry, and coherent anti-stroke Raman scattering and Acoustic techniques. Nevertheless, the low cost, and simplicity of thermocouples make them the most widely used sensors. A thermocouple sensor measures the temperature of the thermocouple junction itself, which of course is not what we usually want to know. Rather, we wish to determine the temperature of the body in which the thermocouple is embedded. In addition to that, their accuracy is affected by radiation and conduction errors. At high temperature radiation error increases enormously. For accurate measurement bead of thermocouple should be as small as possible. If a sheathed thermocouple is used then metal protective sheath around the thermocouple should be also thin. Thermocouples are thus fragile and unsuitable for extreme environments at high temperature and supersonic speeds [1]. Infrared thermometry has high initial cost and accuracy is affected by surface emissivity, reflections, fluorescence, absorption and scattering of signal. Other techniques like Coherent Anti-Stroke-Raman Scattering and Acoustic techniques are very costly and complex in operation [2].

It is thus clear that a new technique is required to measure high temperature at supersonic speed. Development of a gas dynamic probe to measure temperature in such high speed environment is discussed in this paper.

2. Experience to Date

Perry L. Blackshear Jr. [3] developed a pneumatic probe which consisted of two sonic-flow type orifices in series. Comparison with iron-constantan thermocouple over a progressive variation of temperature from 400K to 650K gave excellent agreement. In comparison with Sodium D-Line method within the range of 1800K to 2222K it indicated a difference in temperature of less than 2% at equilibrium conditions and 6% at non-equilibrium conditions.

Marvin D. Scadron [4] constructed a pneumatic probe based on continuity of mass flow through a subsonic orifice plate and sonic converging-diverging nozzle separated by a cooling chamber. Comparison was made with a rake of radiation shielded thermocouples at maximum temperature of 1100K and indicated Mach number up to 0.8. Maximum error in measurement was observed to be 1.5%.

C. Dewey Havill& L. Stewart Rolls [5] developed a temperature-measuring system using two sonic-flow orifices in series to measure the exhaust gas temperature of an afterburning jet engine. Comparison was made with shielded chromel – alumel thermocouple and measurement of exhaust temperatures of 1700 K was obtained during afterburner operation in flight.

Frederick S. Simmons & George E. Glawe [6] developed a pneumatic temperature probe. Temperatures between 880 K and 2200 K were measured with an experimental probe in hydrocarbon combustion exhaust gases and the results compared were in good agreement with thermocouple pyrometers and line-reversal pyrometer.

Yang, X.L, Miller, and Hodson, H.P [2] measured the temperature using the choked nozzle in the hot flow and the orifice as the second restriction. They calibrated the probe in the range of 300 K to 900 K. They obtained the accuracy of 1% at 2000 K.

MichelaMassini, Robert J. Miller & Howard P. Hodson [7] developed a probe which consisted of a choked nozzle located in the flow and a downstream system including a cooler, flow-meter and a valve. The probe prototype was tested up to 900K and showed an accuracy of ±6K.

K. Sathiyamoorthy [8] developed a low cost, simple in construction gas dynamic probe for total temperature measurement of hot gases. Experiments were conducted by varying P_{01} from 4 to 6 bar, T_{01} from 300 K to 900 K and Mach number from 0.4 to 1.8. It was observed that the variation in probe constant C with Mach number and total pressure P_{01} for a particular total temperature T_{01} was less than 1%.

3. Gas Dynamic Probe

This section describes the aero-thermodynamic design of gas dynamic probe for total temperature measurement of high temperature ($T_{o\infty}$) flow at Mach 2. The gas dynamic probe measures the total temperature indirectly using gas dynamic relations. Schematic diagram of gas dynamic probe is depicted in figure 1. It consists of two choked nozzles in series. Since the incoming flow is supersonic at Mach 2, a bow shock is formed at upstream of the probe and makes the incoming flow to the probe subsonic. It is accelerated by the first nozzle so that the flow gets choked at its throat. Since the flow gets choked at throat of first nozzle, the flow rate is a function of total temperature, total pressure and specific heat ratio and gas constant. The hot flow is cooled after passing through the first nozzle, in a small settling chamber, using water based heat exchanger. The flow temperature is reduced to 380-430 K in the settling chamber. The flow from settling chamber is then accelerated to Mach 1 in second nozzle, to choke at its exit. Second nozzle is a convergent nozzle type with throat diameter 2.1 mm. Total temperature is measured in the settling chamber using a K-type thermocouple. Also a static pressure tapping is made in the settling chamber. When the shut-off valve (SOV) is open it gives static pressure (P_2) value of the flow in the settling chamber. But when SOV closes the flow it reads total pressure (P_{o1}) in the settling chamber is obtained. Two configurations of gas dynamic probe have been designed. Congiuration-1 has a convergent-nozzle as the first nozzle followed by the settling chamber, while the configuration-2 has a convergent-divergent nozzle as the first nozzle followed by the settling chamber. Throat diameters of first nozzles in both the configuration is same and equals to 2 mm. In both the configurations second nozzle is convergent type having same dimensions (having throat diameter of 2.1 mm). The miniature settling chamber has diameter of 4 mm. The first nozzle for both the configurations is shown in figure 2. The total pressure loss due to formation of shock at the exit of the first nozzle of second configuration will be lesser than that in configuration-1. This is because of sudden increase in are at nozzle exit in configuration-1. Whereas, in configuration-2 there is gradual increase in are after throat to settling chamber. Nozzles are made of SS304 material using electric discharge machining. Figure 3 shows the fabricated gas dynamic probe (configuration-2) along with its major components, cooling water path and settling chamber measurements. Figure 4 shows the flow path of gas and cooling water. Blue and red arrows depict the cooling water and gas flow paths respectively. The cooling water first goes to the first nozzle followed by the settling chamber. The heat exchanger is a combination of parallel and counter flow types. The cooling water flows opposite to the gas flow to reach first nozzle. After that it turns and goes to settling chamber, parallel to the gas flow as shown in figure 4.

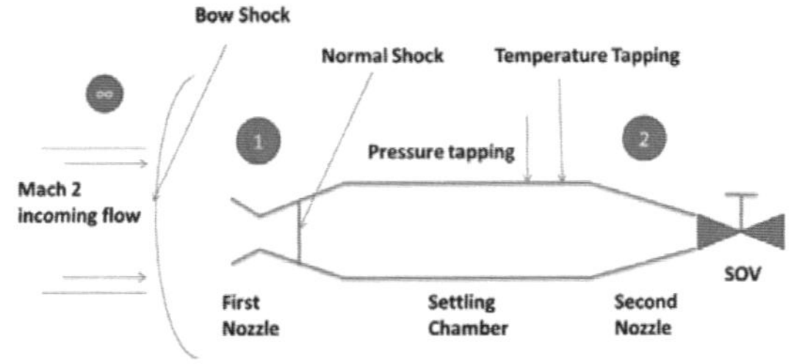

Figure 1. Schematic of the Gas Dynamic Probe configurations

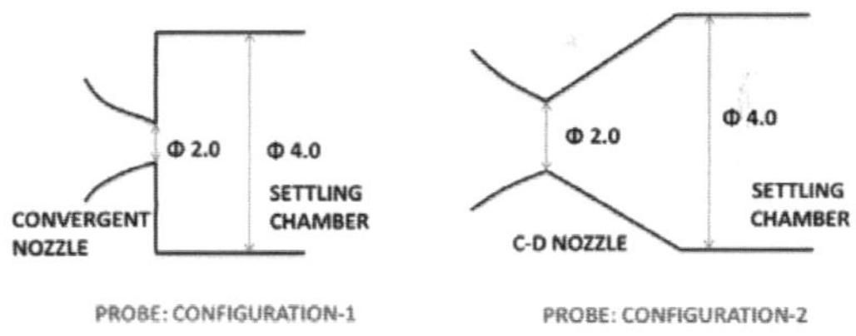

Figure 2. First nozzle of two probe

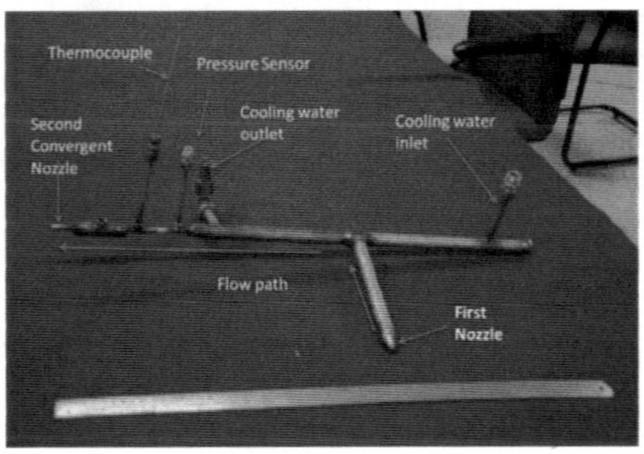

Figure 3. Gas Dynamic Probe: Configuration-2

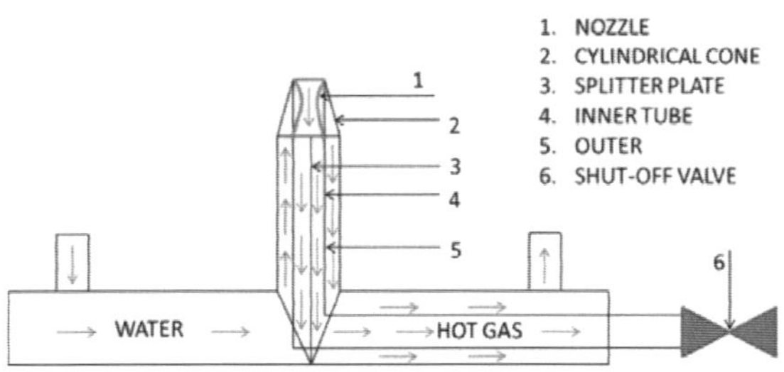

Figure 4. Flow paths of cooling water and hot gas

IOP Conf. Series: Materials Science and Engineering **234** (2017) 012004 doi:10.1088/1757-899X/234/1/012004

Under steady state condition and from the continuity equation, it can be stated that the mass flow rate through the first nozzle should be equal to the mass flow rate through the second nozzle. Suffices ∞, 1 and 2 correspond to aero-thermodynamic and geometrical properties of free-stream incoming air, first nozzle (before settling chamber) and second nozzle (after settling chamber) respectively. Mass flow rate through the first nozzle can be calculated using choking mass flow parameter equation as below:

$$\dot{m}_1 = A_1^* P_{o1} \sqrt{\frac{\gamma_1}{RT_{o1}}} \left(\frac{\gamma_1 + 1}{2}\right)^{-\frac{\gamma_1+1}{2(\gamma_1-1)}} \tag{1}$$

Mass flow rate through the second nozzle is similarly given as below:

$$\dot{m}_2 = A_2^* P_{o2} \sqrt{\frac{\gamma_2}{RT_{o2}}} \left(\frac{\gamma_2 + 1}{2}\right)^{-\frac{\gamma_2+1}{2(\gamma_2-1)}} \tag{2}$$

Probe constant is defined as the ratio of mass flow rate through second nozzle and mass flow rate through the first nozzle. Ideally the value of probe constant should be unity.

$$C = \frac{\dot{m}_2}{\dot{m}_1} \tag{3}$$

Equations 1, 2 and 3 can be re-written to find the inlet total temperature To1, as below:

$$T_{o1} = T_{o2} \frac{\gamma_1}{\gamma_2} \left[C \frac{A_1^* P_{o1}}{A_2^* P_{o2}} \right]^2 \frac{\left(\frac{\gamma_1+1}{2}\right)^{-\frac{\gamma_1+1}{2(\gamma_1-1)}}}{\left(\frac{\gamma_2+1}{2}\right)^{-\frac{\gamma_2+1}{2(\gamma_2-1)}}} \tag{4}$$

Since the flow across a shock wave is isenthalpic, total temperature across the shock wave may be assumed constant. Hence measure of T_{o1} gives incoming flow total temperature $T_{o\infty}$.

4. Experimental Setup

Experiments have been conducted at High Speed Combustor test Facility (HSCTF), Propulsion Division, CSIR-National Aerospace Laboratories, Bangalore. Figure 5 and 6 show the test rig schematic and test rig respectively.. The clean and pressurized air is supplied by a centralized air storage facility. A gate valve and an Electro-pneumatic control valve are used to get the air at required mass flow rate and pressure. Pressures are measured at required points in the air supply system and mass flow rate is measured using orifice meters. The pre-heaters are kerosene based and result in vitiated type air heating. The air is heated in two stages in pre-heaters. The pre-heater-1 is a can combustor which uses kerosene as fuel to increase the temperature from 300K to 1000K. Pre-heater-2 is afterburner based which increases the temperature from 1000K to required value. The settling chamber after the pre heaters is made out of carbon steel and the walls are insulated with ceramic layers to absorb the thermal loads to protect the ducts from deterioration. Flow is accelerated to Mach 2.0 using C-D uniform flow nozzle at the exit of the settling chamber. The nozzle is 2-dimensional. Total pressure of the hot air ($P_{o\infty}$) and total temperature ($T_{o\infty}$) are measured in the settling chamber. Static pressure is measured at the nozzle exit. Using this static pressure and settling chamber total pressure, facility nozzle exit Mach number is obtained.

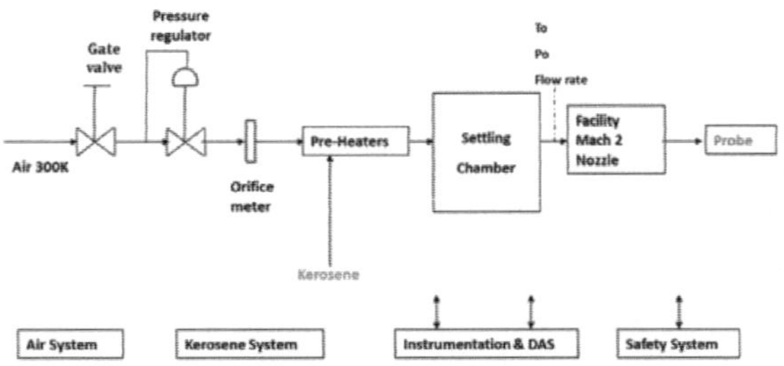

Figure 5. Test Rig Schematic

Figure 6. Test Rig

IOP Conf. Series: Materials Science and Engineering **234** (2017) 012004 doi:10.1088/1757-899X/234/1/012004

The probe is fixed at the exit of the facility of the test rig such that the upstream nozzle of the probe is inserted directly in the hot exhaust stream (shown in figure 7). The sequence of the experiment is as follows. The air is allowed through the test rig using gate valve and control valve at desired pressure and mass flow rate. Then pre-determined value of kerosene is injected in the pre heater to achieve the required temperature. Once the simulation parameters are achieved, the output of the sensors is recorded through the data acquisition system for the open and closed condition of the shut off valve in the gas dynamic probe. Data is recorded after reaching steady state. The pressure is measured using PCB make pressure transmitters and temperature is measured using R-type thermocouple. The pressure and temperature data is collected and stored by Data Acquisition System (DAS).

The total pressure (P_{o1}) of the flow approaching the first nozzle is measured by closing the shut off valve in the probe. Whereas the total temperature (T_{o1}) is taken equal to incoming flow temperature ($T_{o\infty}$). Using these total temperature and total pressure, mass flow rate through the first nozzle is calculated. Under the open condition of the probe shut off valve, the pressure measured is equal to static pressure (P_2) in the settling chamber. The total temperature obtained (T_{o2}) is around 380-430 K due to cooling by water. Using area ratio relationship and known temperature, Mach number at the settling chamber is calculated. Using this Mach number and static pressure (P_2), the total pressure of flow (P_{o2}) of second nozzle is obtained. Hence, using T_{o2} and P_{o2}, mass flow rate through second nozzle is obtained.

Figure 7. Gas Dynamic probe assembled to the Facility Nozzle

5. Results and Discussions

Mass flow rates through first nozzle and second nozzle are calculated using Eq. 1 and Eq.2 respectively. Specific heat ratios in both cases are calculated based on temperature at that condition using NASA SP-273 program. Probe constant is the ratio of mass flow rate through first nozzle and mass flow rate through second nozzle (Eq.3). The experimentally obtained values of probe constant for both the configurations are plotted below with respect to incoming flow total temperature ($T_{o\infty}$) and total pressure ($P_{o\infty}$).

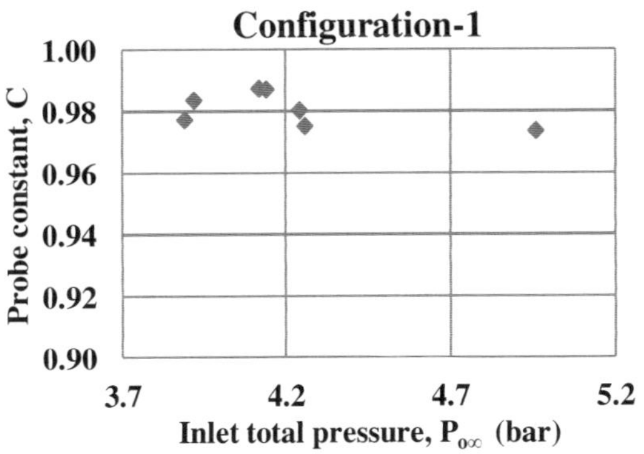

Figure 8. Probe constant vs inlet total pressure

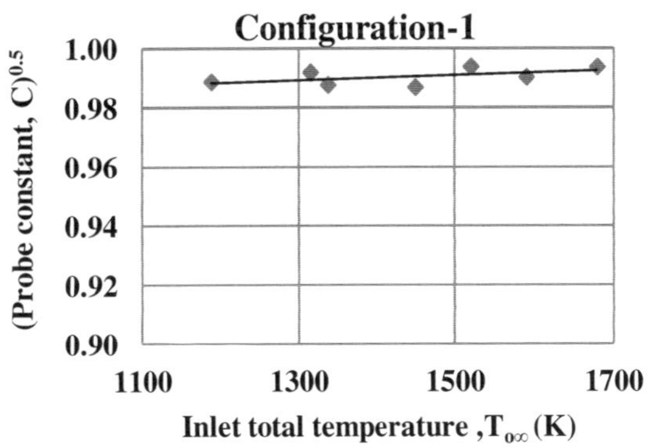

Figure 9. (Probe constant) $^{0.5}$ vs inlet total temperature

IOP Conf. Series: Materials Science and Engineering **234** (2017) 012004 doi:10.1088/1757-899X/234/1/012004

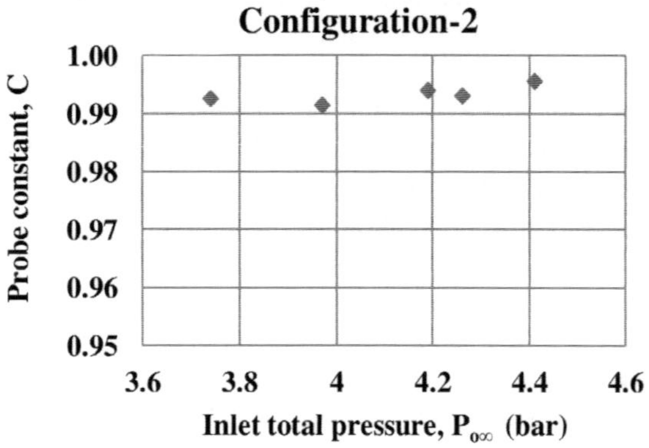

Figure 10. Probe constant vs inlet total pressure

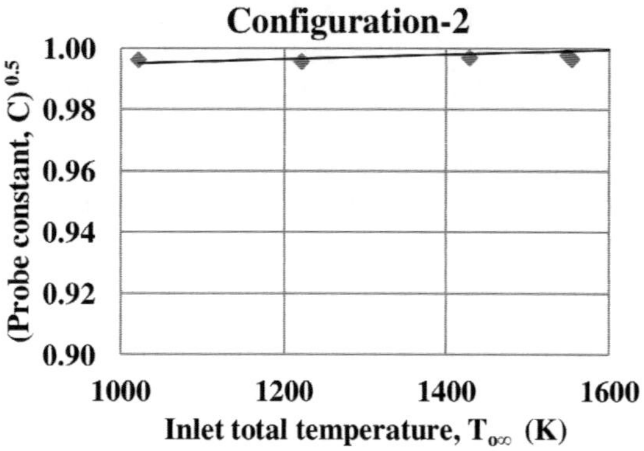

Figure 11. (Probe constant) $^{0.5}$ vs inlet total temperature

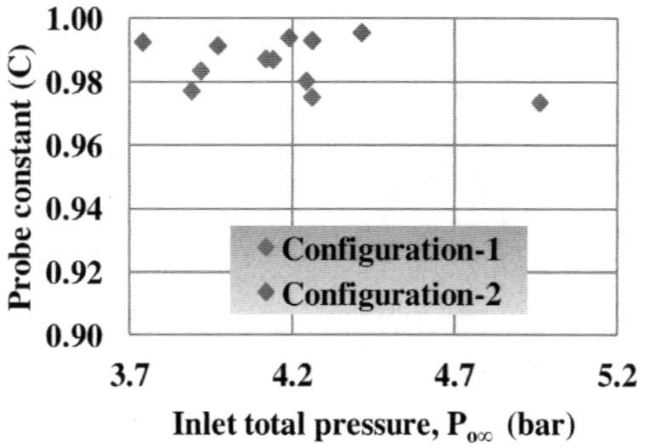

Figure 12. Probe constant vs inlet total pressure

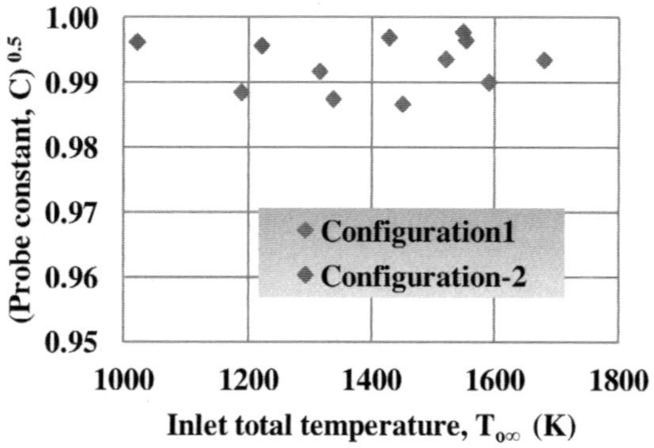

Figure 13. (Probe constant) $^{0.5}$ vs inlet total temperature

Configuration-1 has a convergent nozzle followed by settling chamber. Figure 8 shows the variation of probe constant against incoming flow total pressure for this configuration. It is observed that probe constant varies between 0.974 and 0.987. Figure 9 shows the variation square root of probe constant against total temperature of incoming hot flow for configuration-1. It is observed that square root of probe constant increases linearly with respect to total temperature of incoming flow. On observation of Eq.1, 2 and 3 it is evident that probe constant is proportional to first nozzle throat area. First nozzle sees high temperature gas, whereas the second nozzle gets incoming flow at relatively low temperature. As the diameter of throat of first nozzle increase linearly with temperature, hence its throat area increases proportional to the square of temperature, results in linear increase of probe constant.

Configuration-2 has convergent divergent nozzle followed by settling chamber. Figure 10 shows the variation of probe constant against incoming flow total pressure for this configuration. It is observed that probe constant varies between 0.991 and 0.996. Figure 11 shows the variation of square root of probe constant against total temperature of incoming hot flow for this configuration. It is evident that probe constant increases proportional to square of incoming total temperature.

Figure12 and 13 show comparison between two configurations. Configuration-1 has a convergent nozzle followed by settling chamber. There is a sudden increase in area at the junction of convergent nozzle and settling chamber. Whereas configuration-2 has a convergent-divergent nozzle followed by settling camber. Exit area of the nozzle is equal to the area of the settling chamber. Hence, there is smooth and gradual transition of flow. So the total pressure loss is higher in configuration 1 than configuration 2. This is the reason for higher probe constant values for configuration -2 than that of configuration-1.

6. Conclusions

A water cooled gas dynamic probe was designed and calibrated for total temperature measurement of high speed (Mach 2) and high temperature gases using two choked nozzle as restrictions. Probe constant was experimentally obtained for two configurations of probe by varying incoming flow total temperature and total pressure. Probe constant approximately varies proportional to the square of incoming flow total temperature. First restriction with C-D nozzle gave higher probe constant value than convergent nozzle.

Acknowledgements

Authors wish to express their sincere thanks to the Director, CSIR-NAL Bangalore and Head, Propulsion for providing all the necessary support.

Nomenclature

$\dot{m}_1$: Mass flow rate through first nozzle
$\dot{m}_2$: Mass flow rate through second nozzle
A_1^* : Throat area of first nozzle
A_2^* : Throat area of second nozzle
P_{o1}: Total pressure at first nozzle (SOV closed)
P_2 : Static pressure at second nozzle (SOV open)
P_{o2}: Total pressure at second nozzle (SOV open)
$P_{o\infty}$: Total pressure of incoming flow to the probe
T_{o1}: Total temperature at first nozzle (SOV closed)
T_{o2}: Total temperature at second nozzle (SOV open)
$T_{o\infty}$: Total temperature of incoming flow to the probe
γ_1 : Specific heat ratio at first nozzle
γ_2 : Specific heat ratio at second nozzle
C : Probe constant

References

[1] Manik and Brown, " Temperature: Its Measurement and Control in Science and Industry", American Institute of Physics, New York, 1992, p 1244-1249.
[2] X. L. Yang, R. J. Miller and H. P. Hodson, "A New Probe for the Measurement of High-Temperature Gases", the 16th symposium on measuring techniques in transonic and supersonic flow in cascades and turbo machines, Cambrige U.K, 2002.
[3] Blackshear and Perry L, "Sonic flow orifice temperature probe for high gas temperature measurement," NACA TN 2167, 1950.
[4] Scadron Marvin D, "Analysis of a Pneumatic Probe for Measuring Exhaust Gas temperature with some preliminary results", NACA RM E52A11, 1952.
[5] Havill C, Dewey and Rolls L Stewart, "A sonic flow orifice probe for the in-flight measurement of temperature profiles of a jet engine exhaust with afterburning", NACA TN 3714, 1956.
[6] Simmons, Frederick S and Gwale, "Theory and design of a pneumatic probe and experimental results obtained in a high temperature gas stream" NACA TN 3893, 1957.
[7] Michela Massini, Robert J. Miller, and Howard P. Hodson, "A New Intermittent Aspirated Probe for the Measurement of Stagnation Quantities in High Temperature Gases", J. Turbomach Oct 2010 Volume133, Issue4, 041022(6pages).
[8] K. Sathiyamoorthy, et al., "Development of Gas Dynamic Probe for Total Temperature Measurement in gases", ASET 2011- National Conference on "Emerging Trends in Propulsion Technology".

IOP Conf. Series: Materials Science and Engineering **234** (2017) 012005 doi:10.1088/1757-899X/234/1/012005

Fuel Effective Photonic Propulsion

N Rajalakshmi[1] **and S Srivarshini**[2]

[1] Department of Electrical and Electronics, Sri Sairam Institute of Technology, Chennai, India
raji.n.narayan@gmail.com
[2] Department of Electrical and Electronics, Sri Sairam Institute of Technology, Chennai, India
ashwanth.sridharan1996@gmail.com

Abstract: With the entry of miniaturization in electronics and ultra-small light-weight materials, energy efficient propulsion techniques for space travel can soon be possible. We need to go for such high speeds so that the generation's time long interstellar missions can be done in incredibly short time. Also renewable energy like sunlight, nuclear energy can be used for propulsion instead of fuel. These propulsion techniques are being worked on currently. The recently proposed photon propulsion concepts are reviewed, that utilize momentum of photons generated by sunlight or onboard photon generators, such as blackbody radiation or lasers, powered by nuclear or solar power. With the understanding of nuclear photonic propulsion, in this paper, a rough estimate of nuclear fuel required to achieve the escape velocity of Earth is done. An overview of the IKAROS space mission for interplanetary travel by JAXA, that was successful in demonstrating that photonic propulsion works and also generated additional solar power on board, is provided; which can be used as a case study. An extension of this idea for interstellar travel, termed as 'Star Shot', aims to send a nanocraft to an exoplanet in the nearest star system, which could be potentially habitable. A brief overview of the idea is presented.

1. Introduction

Mankind has always wanted to go beyond the solar system on an interstellar travel. But the nearest stars to the Sun (Alpha Centauri) are located 4.4 light years away. Using conventional spacecrafts, reaching the interstellar systems would not be possible even in a generation's time. This is because of the slow speeds of our crafts. So, the speed of the craft needs to be increased. According to Dr. Robert Frisbee of NASA's Jet Propulsion Laboratory, the fuel takes up almost $2/3^{rd}$ of the craft's mass and is stopping us from reaching higher speeds. If we can find other alternatives for fuel, science fiction levels of speeds can be attained. Utilization of the energy of **light (Photonic propulsion)** strangely proposes a good solution as it allows very little or no fuel to be carried along with the spacecraft. Photons do not have mass, but they have energy and momentum. When they get reflected off an object, that momentum is transferred into a little push (thrust). The thrust depends only on power and not wavelength. Clearly, these spacecrafts needs to be very small in size and cannot be used for manned missions.

2. Motivation

The present near-earth space endeavors are limited fundamentally by the following factors [1]:
(1) Extremely high construction/operation cost, (2) Fuel limited mission lifetime and operation capability, (3) High environmental impact.

How important is that this problem is addressed? Mankind is one step ahead of exploring potentially habitable planets in the universe. If this problem is overcome, interstellar travel can be possible in unimaginable less time and open new gateways for space exploration.

3. Sunlight for photonic propulsion

This **fuel-independent** technique utilizes momentum of sunlight that has the greatest specific impulse. Specific impulse is a measure of the efficiency of rocket and jet engines. Here, since the output has no (rest) mass and is not expended fuel; if we take the momentum per inertia of the photons, the specific impulse is just the speed of light. The spacecraft that incorporates this kind of propulsion is called solar sail. The sail is made up of or coated with highly reflective material, like aluminium. The coating is made over plastic medium for mechanical support. "Fig.1" shows the various forces involved in the momentum transfer from photon to the sail. The force on the sail ("Fig.1") is in accordance with the inverse square law. It depends inversely on the square of the distance of the sail from the Sun. Force on the sail, $F_{sail} \alpha \frac{A\cos\theta}{r^2}$ (1)

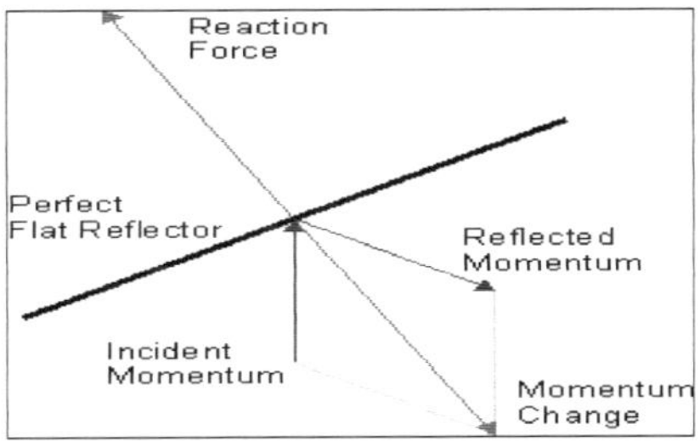

Figure 1. Force on sail due to photon repulsion

3.1. Sail parameters [2],[3]:

- Sail loading σ is the areal density: $\sigma = \frac{massofsail}{sailarea}$ (2)
- Characteristic acceleration a_c is related to σ as $a_c = \frac{efficiency}{\sigma}$ (3)
- Lightness number λ relates the craft's maximum acceleration and the Sun's local acceleration due to gravity (5.93ms^{-2}).

$\lambda = \frac{maximum\ acceleration\ of\ craft}{5.93}$ (4)

There are different types of solar sails based on their structure and appearance whose sail parameters are given in the Table 1.

Table 1. σ, a_c, λ values for various types of solar sails

Type	σ(g/m^2)	a_c(mm/s^2)	Λ
Square sail [4]	5.27	1.56	0.26
Helio-gyro [5]	18.3	0.5	0.03

4. Laser For Photonic Propulsion

The LASER Photonic Propulsion uses direct momentum transfer of photons generated by LASER source. These LASER sources can be Earth-based, placed in space or on board generated using solar or nuclear power. The craft is first suspended in space using a rocket, and then a powerful LASER beam is directed at it from the Earth to propel it forward. It requires beaming of laser photons over astronomical distances to the craft, amplification using two highly reflective mirrors [6], and reflection that pushes the craft. "Fig. 2" shows the schematic of LASER propulsion.

4.1. Challenges and overcoming:

IOP Conf. Series: Materials Science and Engineering **234** (2017) 012005 doi:10.1088/1757-899X/234/1/012005

- *It needs ultra-large, thin space optics*[7]:
  A new emerging technology to fabricate ultra large adaptive space mirrors using the **Photonic Muscle** has been proposed [8] that allows shaping to smaller sizes.
- *High powered LASER:*
  The LASER power required for an **interstellar** mission could be reduced from 64GW to **448 MW** by the use of a dielectric sail [9] with high emissivity and low absorption.
- *Large thrust:*
  Using Photonic Laser Thrusters (PLT), Direct energy (DE) can be used for momentum beaming [2]. Optical resonator can be used for LASER propulsion [10]. Thin Disk Laser can produce PLT thrust of 5mN [11].

The idea of 'Starwisp' proposed by Robert.L.Forward aims at using MASER placed on the Moon[12] that will guide a spacecraft weighing only a few kilograms [13].

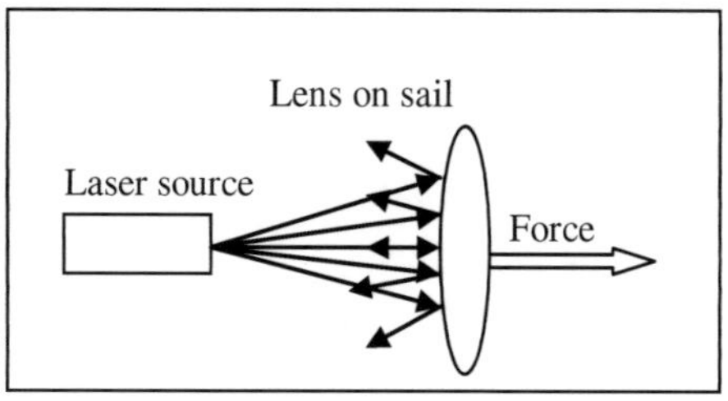

Figure 2. LASER Propelled Light Schematic

5. Nuclear Photonic Propulsion

Nuclear energy is converted into blackbody radiation that is similar to photons (EM-waves) that can be used for propulsion. Such a craft is nuclear photonic rocket. Since it requires some amount of fuel to be taken along with the craft to produce power for thrusting, this method fuel-dependent [14].

5.1. Power requirement calculation

The power per thrust required for a perfectly collimated output beam is 300 MW/N (1N of thrust requires 300 MW of energy [15].)

Example to compute T/W value:

The Russian-made RD-180 rocket engine produces 3,820 kN of sea-level thrust and has a dry mass of 5,307 kg. Using the Earth acceleration due to gravity of 9.807 m/s², the sea-level thrust-to-weight ratio is computed as follows [16]: $\frac{T}{W} = \frac{3820\ kN}{5307 kg * 9.807\ m/s^2} = 73.40$. This value for jet and rocket engines vary from 1.8 to 180 that are used for propelling rockets. For the aircraft having least T/W ratio (RD-0410 nuclear rocket engine), it requires 35.2 kN of thrust for mass of 2000 kg. For photonic sails, the weight is only in grams or a few kilograms, say 5 kg. This requires: $\frac{35.2\ kN * 5\ kg}{2000\ kg}$ =88 N of thrust, that would be provided by: $\frac{300\ MW * 88\ N}{1\ N}$ = 26400 MW = **26.4 GW of power**.

5.2. Fuel requirement calculation

E_{ph}-Photon energy available for propulsion
M -Mass of rocket
α -Mass ratio of fuel to rocket (α<1)
c -Speed of light
δ -propulsion system energy to photon energy conversion efficiency (δ<1)
γ - Fuel mass to propulsion system energy conversion efficiency
T_{ph}-Total photon thrust
V_{max}-Maximum attainable velocity

Here, αM the total mass of the fuel. Considering the above notations, [7]

$E_{ph} = Mc^2\ \alpha\ \delta\ \gamma$ (5)

$T_{ph} = \frac{E_{ph}}{c}$ (6)

$V_{max} \approx \frac{T_{ph}}{M} = \alpha\ \delta\ \gamma\ c$ (7)

We know that T_{ph} is 88N and mass is approximately 5kg.

So, $V_{max} \approx \frac{T_{ph}}{M} = \frac{88}{5*9.8} = 1.79$ m/s (using eqn. 7)

$E_{ph} = T_{ph}c = 88*3*10^8 = 2.64*10^{10} = 26400$MJ (using eqn.6)

Assuming 10% of fuel is effectively converted into propulsion energy, $\gamma = \frac{10}{100} = 0.1$;

$\alpha = \frac{E_{ph}}{Mc^2\delta\gamma} = \frac{26400\ MJ}{5*9.8\ kg*(3*10^8)^2 m/s*0.1*\delta} = 1.197*10^{-7}$ (for δ=0.5) (using eqn.5)

Thus, $1.197*10^{-7}*5*9.8 = 5.86*10^{-6}$ kg = **$5.86*10^{-3}$g of fuel is required.**

5.3. Comparison with conventional propulsion

It requires 35.2 kN of thrust for mass of 2000 kg for conventional rockets. That would be provided by:

$\frac{300\ MW*35200\ N}{1\ N} = 10500$ GW = **10.5 TW of power**.

$V_{max} \approx \frac{T_{ph}}{M} = \frac{35200}{2000*9.8} = 1.79$ m/s (using eqn.7)

$E_{ph} = T_{ph}c = 35200*3*10^8 = 105.6*10^{11}$ J (using eqn.6)

Assuming 10% of fuel is effectively converted into propulsion energy, $\gamma = \frac{10}{100} = 0.1$

$\alpha = \frac{E_{ph}}{Mc^2\delta\gamma} = \frac{105.6*10^{11}}{2000*9.8\ kg*(3*10^8)^2 m/s*0.1*\delta} = 1.197*10^{-7}$ (for δ=0.5) (using eqn.5)

Thus, $1.197*10^{-7}*2000*9.8 = 2.34*10^{-3}$ kg = **2.34g of fuel is required**.

5.4. Results:

It is seen that the fuel requirement for nuclear photonic propulsion is lesser than that of conventional methods. **Weight of fuel saved is 2.33414 grams.**

A broad comparison between the conventional and photonic propulsion techniques is given in Table.2.

Table 2. Comparison of conventional and photonic propulsion

Conventional propulsion	Photonic propulsion
1. Fuel takes most of the craft weight	**1.** Can be completely fuel-independent.
2. For manned missions.	**2.** For unmanned missions.
3. 3% of the speed of light can be achieved.	**3.** 10-30% of that of light can be achieved.
4. Interstellar missions are impossible within a generation's time.	**4.** Interstellar missions are possible within a generation's time.
5. Uses non-renewable sources of energy.	**5.** Uses renewable sources of energy.

6. IKAROS Mission Overview

The IKAROS (Interplanetary Kite craft Accelerated by Radiation Of Sun) is a Japanese spacecraft, deployed on May 21st, 2010. It is the first spacecraft to successfully demonstrate solar sail technology as the main propulsion. It currently explores interplanetary space in orbit between Venus and Earth.

It successfully completed following missions [17]: deploying a large membrane sail, generating electricity by thin film solar cells, check if photon propulsion works.

6.1. Sail Layout

- *Membrane*: The membrane is square shaped whose diagonal distance is 20cm made of polyimide.

- *Thin film solar cell*: Si solar cells attached to some areas of the membrane generate almost 500W.
- *Steering device*: Placed at the tips of the membrane, they can be used to control the spin direction.
- *Dust counter*: PVDF film is attached for dust counter.
- *Tip mass*: Four small masses (0.5 kg) attached to the tips of the membrane to adjust its centrifugal force. [18]
- *Tether*: The membrane is attached to the main body by tethers.

The **sail layout** is shown in "Fig. 3".

6.2. Propulsion system

The highly reflective material of the craft was used for pushing by sunlight photonic propulsion. Additionally, solar cells made from silica generated 500 W onboard the craft.

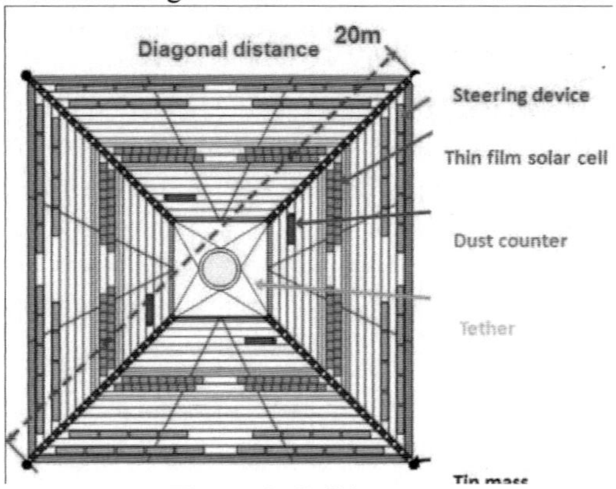

Figure 3. Sail layout

7. Breakthrough 'Starshot' Overview

The 'Starshot' is a breakthrough idea for interstellar travel to nearby galaxies, probably to Alpha Centauri system which is 4.4 light years away [9]. These projects require technological breakthrough, and long-term economic interest and investment [7].

7.1. Sail layout proposal

The sail is proposed to be a **nanocraft** consisting of a single wafer of electronics with cameras, sensors, communication devices, small LASERS for thrusters and a small nuclear battery.

7.2. Propulsion

The craft can be accelerated upto **20% the speed of light using** onboard LASERs and nuclear fuel.

7.3. Reaching the star system and time estimate

This technology is expected to come in 20 years, 20 years to reach Alpha Centauri, 4.4 years to beam data to Earth = **45 to 50 years to get information about the galaxy.**

7.4. Challenges Ahead and Recently Proposed Solutions:

- **The craft should withstand cosmic dust and turbulence**: Three methods are proposed: Route Adjustment, Protective shielding, On-Chip Healing.
- **Miniaturization of electronics**: We cannot reduce the size of electronics beyond a certain limit because quantum tunneling effects come into existence (end of Moore's law) [19]. The base resins for IC is selected by moisture performance and fluidity. [20]
- **Pointing the LASER** array towards the tiny spacecraft is a huge challenge.

8. Conclusion

The power that lies dormant within photons could be mankind's future towards energy efficient flight. Information about other galaxies can be acquired, that is nowhere as good as those acquired by conventional and other observation methods. This technique makes space travel less-dependent on conventional resources by relying more on renewable energy like Sunlight. Thus, if long-term research of photonic propulsion is fueled by constant economic interest and investment, it would take us on the ultimate path to fuel independence for spaceflight.

Reference

[1] Young K. Bae "DE Momentum Beaming (DEMB) for Innovative Spacecraft Maneuvering", *Proc. of the 15th Annu. Directed Energy Symp.*, Nov. 2012, p. 1.
[2] Dandouras, B. Pirard, and J.Y. Prado, "High Performance Solar Sails For Linear Trajectories And Heliostationary Missions", in *Advances in Space Research,* Volume 34, Issue 1, 2004, pp. 198–203,ISSN0273-1177,Available:www.sciencedirect.com/science/article/pii/S0273117704002704
[3] Colin R. McInnes, "Solar sailing: mission applications and engineering challenges", in *Philosophical Transactions A: Mathematical, Physical and Engineering Sciences*, 361 (1813), pp. 2989-3008, ISSN 1364-503X, Available: http://dx.doi.org/10.1098/rsta.2003.1280
[4] "Solar Sail", Available: http://www.theinfolist.com/php/SummaryGet.php?FindGo=photon_sail
[5] W. Keats Wilkie *et al.*, "Heliogyro Solar Sail Research at NASA", in *Advances in Solar Sailing,* pp.631-650, Online ISBN: 978-3-642-34907-2, doi: 10.1007/978-3-642-34907-2 39.
[6] Young Bae. "Photonic Laser Propulsion (PLP): Photon Propulsion Using an Active Resonant Optical Cavity", *AIAA SPACE 2007 Conference & Exposition,* 18-20 Sep, 2007, Copyright © 2007 by the American Institute of Aeronautics and Astronautics, Inc, doi: 10.2514/6.2007-6131.
[7] Young K. Bae," Prospective of photon propulsion for interstellar flight" in *Physics Procedia,* Volume 38, 2012, Pages 253-279, ISSN 1875-3892, doi: 10.1016/j.phpro.2012.08.026, *Proc. of Space Technology and Applications International Forum II*, 2012.
[8] http://www.niac.usra.edu/files/library/meetings/fellows.mar07/1350Ritter.pdf
[9] Geoffrey A. Landis, "Advanced Solar- and Laser-pushed Lightsail Concepts", NASAInstitute for Advanced Concepts, May 31, 1999.
[10] Willy L. Bohn, "Novel aspects in laser propulsion", *Proc. SPIE 7005*, High-Power Laser Ablation VII, 70051C, May 14, 2008 doi:10.1117/12.785634
[11] Bae, Y.K., "Propellantless Spacecraft Formation-Flying and Maneuvering with Photonic Laser Thruster". *NASA Innovative Advanced Concepts (*NIAC*)*, 2014 Symp.
[12] David Shrunk, Burton Sharpe, Bonnie L.Cooper, Madhu Thangavelu, "The Moon: Resources, Future Development, and Settlement", ISBN 978-0-387-73982, p173.
[13] Gregory L. Matloff, "Deep space probes", p58, ISBN: 978-1-4471-3641-5
[14] M. Auweter-Kurtz, C. Bruno, D. Fearn, H. Kurtz, T.J. Lawrence, R.X. Lenard, "IAA Commission III: SG2 - Nuclear Space Power And Propulsion", *technical report, International Academy of Astronautics*, 2008.
[15] Carlo Rubbia (Pavia U.), "Nuclear Space Propulsion With A Pure Electro-Magnetic Thrust", March 11, 2002, CERN-SL-2002-006-EsCT
[16] "Solar Sail", Available: http://www.theinfolist.com/php/SummaryGet.php?FindGo=Thrust_to_weight%20Ratio
[17] Osamu Mori *et al.,* "IKAROS and Solar Power Sail-craft Missions for Outer Planetary Region Exploration", 11th Low Cost Planetary Missions Conference, June 9-11, 2015, abstract.
[18] Tsuda*et al., "*Attitude Determination and Control Operation of IKAROS in Initial On-Orbit Phase", March 2011*, Proc. of The 20th Workshop on JAXA Astrodynamics and Flight Mechanics,* published by Institute of Space and Astronautical Science, Japan Aerospace Exploration Agency (JAXA)(ISAS), Vol: 20, NCID: AA11567475, SHI-NO: AA0065115005, URI: https://repository.exst.jaxa.jp/dspace/handle/a-is/20519
[19] R. W. Keyes, "Physical limits in digital electronics," in *Proceedings of the IEEE*, vol. 63, no. 5, pp.740-767,May1975.doi: 10.1109/PROC.1975.9825
[20] S.Ito et al., "Solid Type Cavity Fill and Under Fill Materials for new IC Packaging Applications", *Proc. of 45th Electronic Components and Technology Conf.,* Las Vegas, 1995, pp. 1217-1222, doi: 10.1109/ECTC.1995.517845

IOP Conf. Series: Materials Science and Engineering **234** (2017) 012006 doi:10.1088/1757-899X/234/1/012006

Numerical Investigations on Aerodynamic Forces of Deformable Foils in Hovering Motions

Yong Zhao[1], Zhen Yin[2], Xiaohui Su[3], Jiantao Zhang[3] and Yuanwei Cao[4]

[1] School of Engineering, Nazarbayev University, 53 Kabanbay batyr Ave., Astana, 010000, Republic of Kazakhstan
[2] School of Energy and Power Engineering, Dalian University of Technology, Dalian, 116024, China
[3] School of Hydraulic Engineering, Dalian University of Technology, Dalian, 116024, China
[4] Institute of Nuclear Engineering, Technical University of Munich, 85747, Garching, Germany

yong.zhao@nu.edu.kz

Abstract. The aerodynamic effects of wing deformation for hover flight are numerically investigated by a two-dimensional finite-volume (FV) Arbitrary Langrangian Eulerian (ALE) Navier-Stokes solver. Two deformation models are employed to study these effects in this paper, which are a full deformation model and a partial deformation one. Attentions are paid to the generation and development of leading edge vortex (LEV) and trailing edge vortex (TEV) which may illustrate the differences of lift force generation mechanisms from those of rigid wings. Moreover, lift coefficient Cl, drag coefficient Cd, and figure of merit, as well as energy consumption in hovering motion for different deformation foil models, are also studied. The results show that the deformed amplitude, 0.1*chord, among the cases simulated is an optimized camber amplitude for full deformation. The results obtained from the partial deformation foil model show that both Cl and Cd decrease with the increase of camber amplitude. It is found that the effect of deformation in the partial deformation model does not enhance lift force due to unfavorable camber. But TEV is significantly changed by the local AOA due to the deformation of the foil. Introduction.

1. Introduction

In the last decade, attention has increasingly been paid to the aerodynamics of deformable wings[1-10] due to the rising popularity of micro air vehicles (MAVs). Researchers have gradually obtained systematic and deep insights into unsteady aerodynamic forces on rigid wings[11-18]. With those insights obtained, further research on flexible wings found in nature and their lift generation mechanisms is expected. However, the structures and materials of flexible wings are found to be so complex that so far one could not yet find simple and accurate models to represent major wing features by measurement data or regression fitted values. Therefore, the prescribed deformation models are adopted in this paper to investigate this problem, which can be considered as semi-rigid wing models because wing deformation is prescribed with simplified harmonic equations during wing flapping based on experimental data. Hovering flight mode is employed where the body is assumed to be fixed in space and the freestream velocity is zero[1,3,11]. As for deformable wings, the hovering motion is considered as a combination of three components: translation, rotation as well as deformation, respectively. And In order to understand the effect of wing flexibility on the aerodynamics of flapping motion, two types of deformable foil models can be used for the purpose of experiment and numerical investigation. They are full deformation foil models[4,19,20] and partial deformation ones.

2. Mathematics

The incompressible unsteady Navier-Stokes governing equations, modified by the Artificial Compressibility Method (ACM) with dual-time steps and Arbitrary Langrangian Eulerian (ALE) Method in non-dimensional vector form are used to simulate the flapping aerodynamics. And the solution methods, including discretization strategy and numerical schemes, as well as model validation for the above ALE governing equations, have been described in detailed by Su et. Al.[21], which will not be repeated in this section.

3. Results and Discussions

In this paper, all 2D foil models mentioned above in section one are used to investigate the effects of foil deformation on lift force generation mechanisms. The shape equations of full and partial deformation foil models are described as follows:

$$\frac{x^2}{a^2}+\frac{y^2}{b^2}=1 \tag{1}$$

where a and b correspond to the major and minor axises of ellipse. x and y mean horizontal and vertical coordinate, respectively. In this paper, the ratio of long axis to short one is set to 30:1.

3.1. Models of the full and partial deformation foils

Translation and rotation motions as well as wingbeat frequency are set to be the same as our previous work for rigid wing investigation[21] for the purpose of comparison. As for the full deformation foil, the deformation of camber is introduced to the central line of the foil and the whole foil deforms accordingly (See figure. 1(a)). In general, two quadratic functions are adopted to represent the camber shape of the central line of the foil for the full deformation foil model as follows:

$$\begin{cases} \dfrac{y_c}{c}=\dfrac{m(t)}{p^2}[2p\dfrac{x}{c}-(\dfrac{x}{c})^2] & (0\le\dfrac{x}{c}\le p) \\ \dfrac{y_c}{c}=\dfrac{m(t)}{(1-p)^2}[(1-2p)+2p\dfrac{x}{c}-(\dfrac{x}{c})^2] & (p<\dfrac{x}{c}\le 1) \end{cases} \tag{2}$$

where x means the abscissa of a point on the chord, y_c is the deviation of the mean line, and m the maximum of the deviation which is called the maximum camber, here p is set to 0.5.

As for the partial deformation foil model, which is one-quarter rigid and the rest flexible, only one quadratic function is adopted to represent the camber shape of the mean line as follows:

$$\frac{y_c}{c}=\frac{m(t)}{(1-p)^2}[(\frac{x}{c})^2-2p\frac{x}{c}+p^2] \quad (p<\frac{x}{c}<1) \tag{3}$$

where pc is the value of the camber position along the foil, and $m(t)$ in equation (2) and (3) are described as follows:

$$m(t)=m_0\sin(2\pi f_0 t) \tag{4}$$

IOP Conf. Series: Materials Science and Engineering **234** (2017) 012006 doi:10.1088/1757-899X/234/1/012006

where m_0 is the amplitude of the camber in time series. m_0 is set from 0 (rigid case) to 0.2*chord in the simulations[2,4], which is plotted in figure 1.

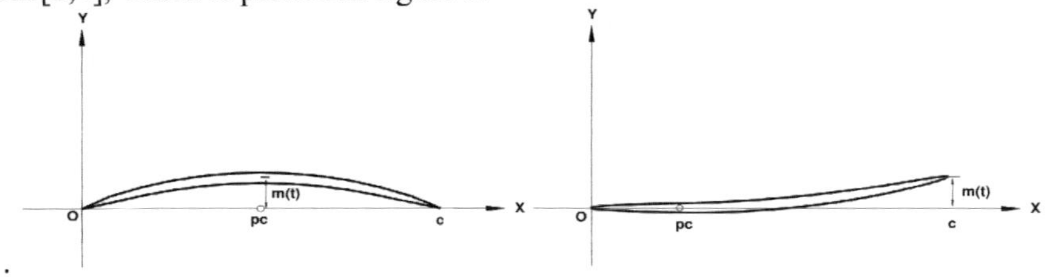

(a) The sketch of full deformation foil model. (b) The sketch of one quarter rigid foil model.

Figure 1. The sketches of full deformation foil model and one quarter rigid foil model.

3.2. Results of the full deformation foil

In this subsection, the differences of aerodynamic forces in hovering motion of the deformable and rigid foils are to be studied. The kinematic parameters of the movement of the hovering foils are set as: $A_0 / c = 3.185$, $f_0 = 0.1$, $\phi = 0$, $\mathrm{Re} = 100$, and camber amplitude, m_0, varies from 0 to 0.2*chord. Based on the kinetic parameters mentioned above, the foil trajectory of the full deformation model is plotted in figure 2.

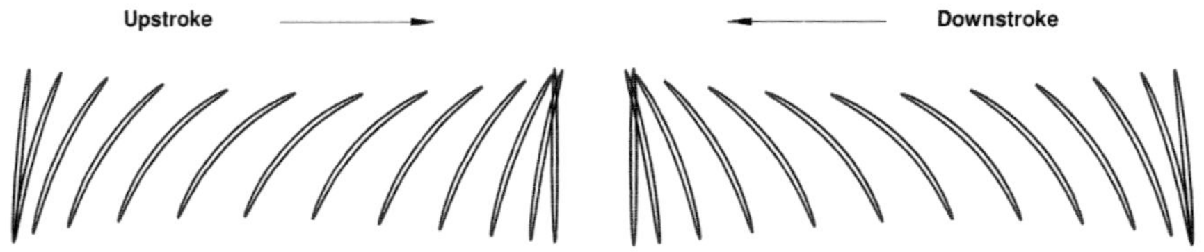

(a) Upstroke for the full deformation model (b) Downstroke for the full deformation model

Figure 2. The foil trajectory of the full deformation model

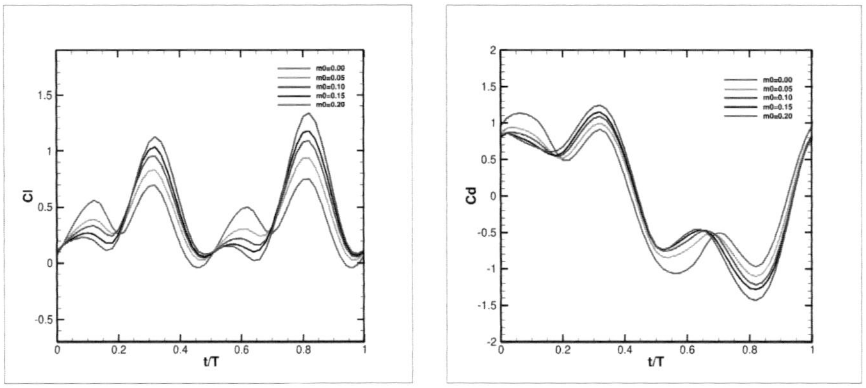

(a) The time-histories of lift coefficients (C_l) (b) The time-histories of drag coefficients (Cd)

Figure 3. The time-histories of lift and drag coefficients for various camber amplitudes in the full deformation model.

Figure 3 gives the time-histories of lift force coefficient (C_l) and drag force coefficient (Cd) of the fully deformable foil in one hovering motion cycle for different camber amplitudes. One can observe clearly that no matter how much camber amplitude is used, there are four crests and troughs for both the Cl and Cd curves. Based on our understanding of the lift force generation mechanisms for rigid wing in hovering motion, the first and the third crests appearing in Figure 3 are generated by the wake capture mechanism[17-18], while the other two crests are generated by the delayed stall mechanism[15].

3.3. Results of the partial deformation foil model

The partial deformation foil model, which was proposed by Miao[22] to explore high lift force generation during insect flight, is investigated in this section by numerical simulations. The kinematic parameters of the hovering motion are same with the full deformations'. The trajectory of hovering motion for partial deformation foil model is plotted in Figure 4.

(a) Upstroke for partial deformation foil model (b) Downstroke for partial deformation foil model

Figure 4. The trajectory of hovering motion for partial deformation foil model.

Figure 5 shows the profiles of lift force coefficient (Cl) and drag force coefficient (Cd) of the partial deformation foil model in one hovering motion cycle for different camber amplitudes. Figure. 5 shows that deformation leads to negative Cl difference values during the hovering motion cycle. This means that the deformation of the wing results in smaller lift force in the partial deformation foil model.

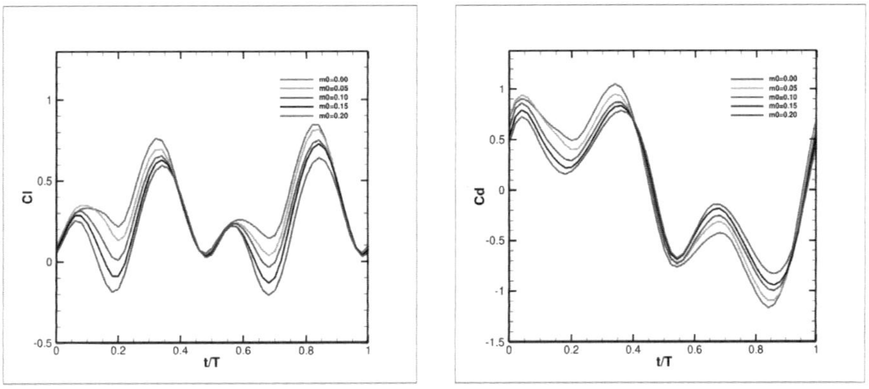

(a) The time-histories of lift coefficients (Cl) (b) The time-histories of drag coefficients (Cd)

Figure 5. The time-histories of lift and drag coefficients for various camber amplitudes in partial deformation foil model test.

3.4. Energy consumptions for different models

Energy consumption is one of the most important considerations in the aerodynamic design of MAVs. The method for calculating energy consumption in the rigid model is to obtain the work which supports the foil to complete the hovering action[23] by using the formulae:

$$P = \int_0^1 C_d\left(\frac{t}{T}\right) * \vec{U}\left(\frac{t}{T}\right) * d\left(\frac{t}{T}\right) \tag{5}$$

where $\vec{U}$ is the dimensionless velocity vector of the rigid foil and $\vec{U}(t/T) = \vec{U}_{real}(t/T)/U_{\max}$, here $\vec{U}_{real}$ is the real velocity vector and . $U_{\max} = \pi f_0 A_0$ is drag force coefficient and T is the period of hovering motion. However, the above methodology needs to be modified as the effect of deformation is included in studying hovering motion. In this paper, we adopt the following formula to investigate the work required within T:

$$P = \int_0^1 \oint_l (\vec{U}(\frac{t}{T}) \cdot \vec{n}(\frac{t}{T})) p(\frac{t}{T}) d\vec{l}(\frac{t}{T}) d(\frac{t}{T}) \tag{6}$$

where $\vec{U}(t/T)$ is the same dimensionless velocity vector as above; $\vec{n}$ is the normal unit vector of a moving boundary element surrounding the node; $p(t/T)$ is dimensionless pressure at the node and $\vec{l}$ the dimensionless boundary element's length. In order to study energy consumption in hovering motion for different deformation foil models, we use the Figure-of-Merit M adopted by Granlund, Ol and Bernal[24], which is the ratio of ideal power to actual power.

$$M = \frac{\overline{C_l}^{2/3}}{2P(2A_0/c)^{1/2}} \tag{7}$$

All the results obtained by numerical simulations, which includes C_l, C_d, C_l/C_d, work, and Figure-of-Merit are plotted in figure. 6 and figure. 6 shows that both Cl and Cd increase with camber amplitude in the full deformation foil model, while the tendency is totally reverse in the partial deformation one. Although work increases with camber amplitude in the full deformation foil model, there exists an optimized value for Figure-of-Merit, 0.3479. The results indicate that larger camber amplitude may not lead to better performance for the full deformation foil model. Among the cases we simulated here, it seems the optimized camber amplitude is 0.1*chord. Compared with the results of fixed chord case (m=0.0), the Figure-of-Merit does not change too much in various chord cases (m>0), for examples, from 0.3278 (m=0.0) to 0.3479 (m=0.1) in full deformation foil model, from 0.3309(m=0.0) to 0.3433(m=0.1) in partial deformation foil model.

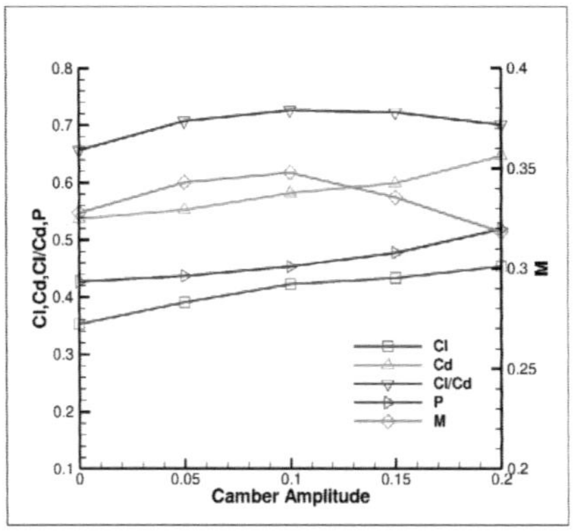

(a) Energy consumption profiles for Full deformation foil model

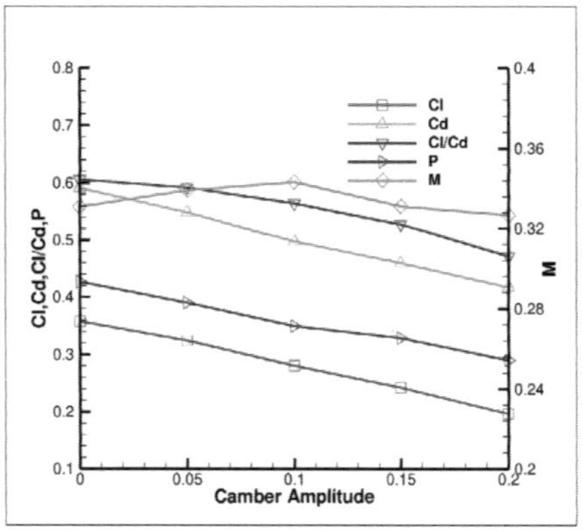

(b) Energy consumption profiles for partial deformation model

Figure 6. Energy consumption profiles for full and partial deformation foil models.

4. Conclusions

In this paper, the effects of deformation on lift force generation mechanisms of wings in hovering flight are investigated. Two foil deformation models, i.e. the full deformation and partial deformation models, are adopted in the research work. The prescribed deformation changes foil morphing during hovering motion in both camber and angle of incidence. The effects of camber amplitude and rotation location on their aerodynamic performances and flow structures are studied in details. The numerical results obtained show that the deformation of foils indeed affects their unsteady aerodynamic performances during their hovering motion. Foil morphing due to deformation makes both LEV and TEV generation and development processes, as well as the lift force generation mechanisms different from those of the rigid foil model. For the full deformation foil model, the effect of deformation enhances its lift force during both the wake capture and delayed stall. There is an optimized camber amplitude, which is 0.1*chord among those cases simulated.

The results obtained from the partial deformation foil model show that both C_l and Cd decrease with the increase of camber amplitude. The effect of deformation does not enhance lift force due to unfavorable camber in this model, as we discussed in the paper. But TEV is significantly changed by the local AOA due to the deformation of the foil.

Acknowledgments
This research is supported by The National Natural Science Foundation of China (NNSFC), project No: 11672059. The financial support from NNSFC is gratefully acknowledged. Financial support from the DUT Haitian Scholars program is also acknowledged.

References

[1] Du G., Sun M., *J. Exp. Biol.*, **213**(13): 2273-2283(2010).
[2] Walker S. M., Thomas A. L. R., Taylor G. K., *J. R. Soc., Interface*, **7**, 131–142 (2010).
[3] Du G., Sun M., *Appl. Math. Mech.*, **29**: 731-743(2008).
[4] Du G., Sun M., *J. Theor. Biol.*, **300**: 19-28(2012).
[5] Yokoyama N., Senda K., Iima M., Hirai N., *Phys. Fluids*, **25**(2): 021902(2013).
[6] Young J., Walker S. M., Bomphrey R. J., Taylor G. K., Thomas A. L., *Science*, **325**(5947): 1549-1552(2009).
[7] Heathcote S., Gursul I., *AIAA J.*, **45**(5): 1066-1079(2007).
[8] Vanella M., Fitzgerald T., Preidikman S., Balaras E., Balachandran B., *J. Exp. Biol.*, **212**(1): 95-105(2009).
[9] Kang C., Shyy W., 30th AIAA Applied Aerodynamics Conference. 2012-2763, 25 - 28 June 2012, New Orleans, Louisiana.
[10] Zhao L., Huang Q., Deng X., Sane S. P., *J. R. Soc., Interface*, **7**(44): 485-497(2010).
[11] Wang H., Zeng L., Liu H., Yin C., *J. Exp. Biol*, **206**(4): 745-757(2003).
[12] Weis-Fogh T., *J. Exp. Biol*, **56**(1): 79-104(1972).
[13] Lehmann F. O., Sane S. P., Dickinson M., *J. Exp. Biol*, **208**(16):3075-92(2005).
[14] Sun M., Yu X., *Acta Mech. Sin.*, **19**(2):103-117(2003).
[15] Liu H., Aono H., *Bioinspiration & Biomimetics*, **4**(1):015012(2009).
[16] Shyy W., Liu H., *AIAA J.*, **45**(12):2817-2819(2007).
[17] Lehmann F. O., Sane S. P., Dickinson M., *J. Exp. Biol*, **208**(16):3075-92(2005).
[18] Birch J. M., Dickinson M. H., *J. Exp. Biol*, **206**(13):2257-72(2003).
[19] Vogel S., *J. Exp. Biol*, **46**(3):431-443(1957).
[20] Dudley R., Ellington C. P., *J. Exp. Biol*, **148**(1):53-88(1990).
[21] Su X. H., Cao Y. W., Zhao Y., *Phys. Fluids*, **28**(6):95(2016).
[22] Miao J. M., Ho M. H., *J. Fluids Struct.*, **22**(3):401-419(2006).
[23] Sane S. P., *J. Exp. Biol*, **204**(15):2607-26(2001).
[24] Granlund K, Ol M, Bernal L. *J. Fluids Struct*, 2013, **40**(7):337-355.

IOP Conf. Series: Materials Science and Engineering **234** (2017) 012007 doi:10.1088/1757-899X/234/1/012007

Combined Effect of Surface Roughness and Wake Splitter Plate on the Aerodynamic Characteristics of a Circular Cylinder

Iyer Saisanthosh[1], K Arunkumar[2] , R Ajithkumar[3,5] and A R Srikrishnan[4]

[1] Department of Mechanical Engineering, Amrita School of Engineering, Amritapuri, Amrita Vishwa Vidyapeetham, Amrita University, India
[2] Department of Mechanical Engineering, Amrita School of Engineering, Amritapuri, Amrita Vishwa Vidyapeetham, Amrita University, India
[3] Department of Mechanical Engineering, Amrita School of Engineering, Amritapuri, Amrita Vishwa Vidyapeetham, Amrita University, India
[4] Department of Aerospace Engineering, Amrita School of Engineering, Ettimadai, Amrita Vishwa Vidyapeetham, Amrita University, India
[5] Department of Mechanical Engineering, Amrita School of Engineering, Amritapuri, Amrita Vishwa Vidyapeetham, Amrita University, India

amritanjali.ajith@gmail.com

Abstract. This paper is focussed on numerical investigation of flow around a stationary circular cylinder (diameter, D) with selectively applied surface roughness (roughness strips with thickness ‘k’) in the presence of a wake splitter plate (length, L). The plate leading edge is at a distance of ‘G’ from the cylinder base. For this study, the commercial software ANSYS Fluent is used. Fluid considered is water. Study was conducted the following cases (a) plain cylinder (b) cylinder with surface roughness (without splitter plate) (c) Cylinder with splitter plate (without surface roughness) and (d) cylinder with both roughness and splitter plate employed. The study Reynolds number (based on D) is 17,000 and $k/\delta = 1.25$ (in all cases). Results indicate that, for cylinder with splitter plate (no roughness), lift coefficient gradually drops till G/D=1.5 further to which it sharply increases. Whereas, drag coefficient and Strouhal number undergoes slight reduction till G/D=1.0 and thereafter, gradually increase. Circumferential location of strip (α) does not influence the aerodynamic parameters significantly. With roughness alone, drag is magnified by about 1.5 times and lift, by about 2.7 times that of the respective values of the smooth cylinder. With splitter plate, for roughness applied at all ‘α’ values, drag and lift undergoes substantial reduction with the lowest value attained at G/D=1.0.

1. Introduction

Flow around circular cylinder has been extensively studied due to its application in a wide variety of engineering structures such as chimney stacks, heat exchangers, cooling towers and so on. Due to space limitation, only few selected past studies are reviewed here. Bishop & Hassan [1] have reported measurements on lift and drag forces on a stationary circular cylinder in the Reynolds number range between 3600 and 11,000. Further, some of the early works carried out are cited in the review paper

IOP Conf. Series: Materials Science and Engineering **234** (2017) 012007 doi:10.1088/1757-899X/234/1/012007

by Niemann & Holscher [2]. In the 1966, Gerrad [3] has proposed the underlying mechanism of vortex shedding behind a circular cylinder. Further to this, Achenback [4] has revealed results on the effect of surface roughness (applied on the entire surface) on the cross flow over a circular cylinder by measuring the local surface pressures and wall shear stresses. He found that, higher the surface roughness, earlier will be the transition of flow from laminar to turbulent. Dalton & Xu [5] found that placement of a control cylinder has notable influence on the drag and lift of the primary cylinder. Norberg [6] found that between Reynolds number value of 1600 and 20000, lift (rms) increases 10 times. Li-Chen et al [7] have proposed a new passive jet control method to manipulate the wake vortex shedding process.

Past studies show that introducing a splitter plate in to the near wake of a circular cylinder brings significant changes in the aerodynamic characteristics around it (Dehkordi and Jafari [8]; Assi and Bearman [9]).Furthermore, application of selective roughness on a circular cylinder is found to considerably modify the flow-induced vibration response of an elastically mounted cylinder (Chang et al [10]; Hongrae et al [11]; Hongrae et al [12]). This has naturally aroused a curiosity as to how the flow field and the resultant fluid forces change if they are together employed. The present study is an outcome of this inquisitiveness to investigate the combined influence of splitter plate and selectively applied roughness on a circular cylinder in water flow. To point out further, such a study has not been reported so far in the literature.

In the present study, selective roughness is applied on a circular cylinder using a roughness strip having a width ‘b’ (circumferential coverage=10 deg) and roughness height (k). Besides this, a splitter plate is introduced in to the body wake. Thus, the combined influence of surface roughness and splitter plate on the aerodynamic/hydrodynamic characteristics of a circular cylinder is investigated at a Reynolds number value of 17000. The study configuration is shown in Fig.1.

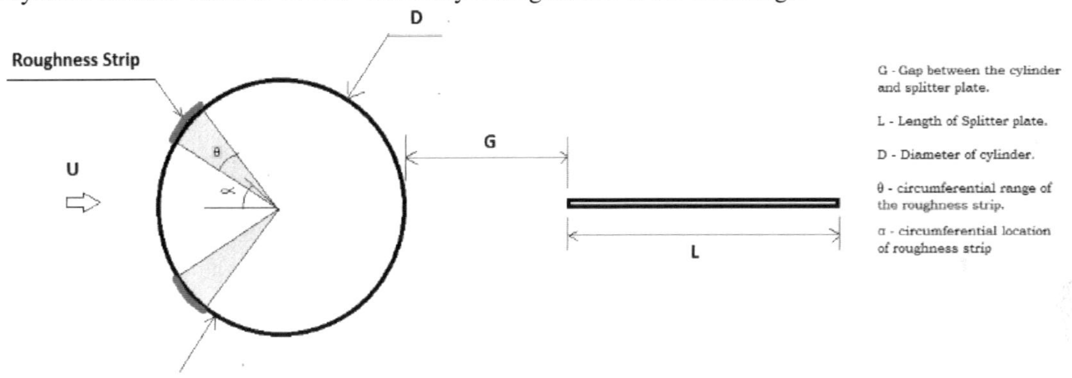

Figure.1: Model geometry

2. Nomenclature

St - Strouhal number
Cpb - Base pressure coefficient
G/D - Gap to diameter ratio
L/D - Length to diameter ratio
α – Roughness location
Cl* - Non dimensional lift coefficient
δ – Boundary layer thickness
k – Roughness height
Cd* - Non dimensional drag coefficient

3. Computational Model and Validation

Numerical simulations in the present study were carried out using ANSYS FLUENT, a finite volume-based commercial CFD tool. Transient formulations of the governing equations for incompressible flow are as follows:

Continuity equation: $\partial \rho / \partial t + \nabla . (\rho V) = 0$ (1)

Momentum equation: $\rho\, (\partial V / \partial t) + \rho V \nabla . V = -\nabla P + \mu \nabla^2 V + \rho g$ (2)

The SST k- ω approach was used for turbulence particularly because of its capability to model flows dominated by separation and recirculation. The domain is discretized using a quadrilateral mesh with adequate resolution of the near-wall regions. A mesh size of 210748 cells was found to be sufficient for a grid-independent solution. The wall y+ is in the range of 2-10. Second order upwind scheme is used for convective and diffusive terms. In order to validate the computational model, the predicted Strouhal number, base pressure coefficient and rms value of lift coefficient were compared with the experimental data reported in the literature at the same Reynolds number. Comparison between the experimental and the present CFD results is summarized in Table 1. The simulation results comply fairly well with the experimentally reported values for the respective parameters.

Table.1: Validation of simulation data with experimental results for smooth cylinder and cylinder with splitter plate

Para-meter	Experiment (Ozono, [13]) –Smooth Cylinder with splitter plate	CFD (Present study)- Smooth cylinder with splitter plate	Experiment (Norberg [6])Smooth Cylinder	CFD (Present study)- Smooth Cylinder
St	0.148	0.1464	0.198	0.196
C_{pb}	0.77	0.75	-------	-------
C_l(rms)	-------	--------	0.446	0.424

The present study is carried out at a Reynolds number (based on D) value of 17,000. The freestream turbulence intensity is 1%. Referring to Fig.1, the roughness strip locations are $\alpha = 0^0$, 9^0, 31^0,65^0 and 75^0. Moreover, for a boundary layer thickness δ, k/δ is kept constant at 1.25 at all strip locations (α). It is pointed out that, the boundary layer thickness vary at each circumferential location and the values calculated at each location is given in Table.2. Boundary layer thickness is calculated based on the modified Pohlhausen method described in Khan et al [14]. The study configurations considered in this series of simulations are as follows (a) stand-alone smooth circular cylinder (b) Circular cylinder with roughness strip at 'α' locations mentioned above (without splitter plate) (c) Cylinder with splitter plate at G/D=0.5, 0.75, 1.0, 1.5 and 2.0, L/D being fixed at 1.0 (without roughness) (d) cylinder with both roughness and splitter employed.

Table. 2: Boundary layer thickness values at various circumferential locations

α	δ (mm)
0^0	0.299215
9^0	0.300321
31^0	0.312966
65^0	0.370096
75^0	0.401496

4. Results and Discussion

Results for the study configurations comprise of drag coefficient, lift coefficient (root mean square value), and the Strouhal number. All these parameters are non-dimensionalized with respect to the respective values for the smooth cylinder and are presented in Figs. 2-4. As seen in Fig.2, the non-dimensional lift coefficient undergoes a reduction as G/D increases from 0.5 to 1.5 further to which it drastically increases to a high value. Non-dimensional Strouhal no. and drag coefficient slightly reduce when G/D is increased from 0.5 to 1.0 and it registers an increasing trend for G/D > 1.0 (Fig.2).

With splitter plate alone, at G/D=1.5, lift reduces substantially to value of about 25% that of the smooth cylinder whereas drag reduces to about 75% that of the smooth cylinder at G/D=1.0.

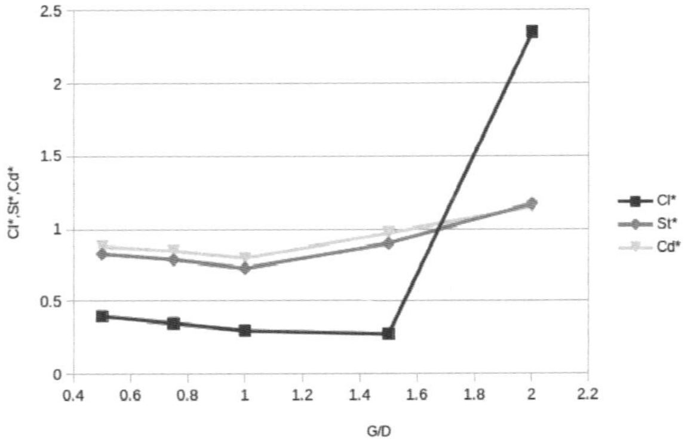

Figure.2: Variation of Cl, Cd and St with G/D ratio (Plain cylinder)

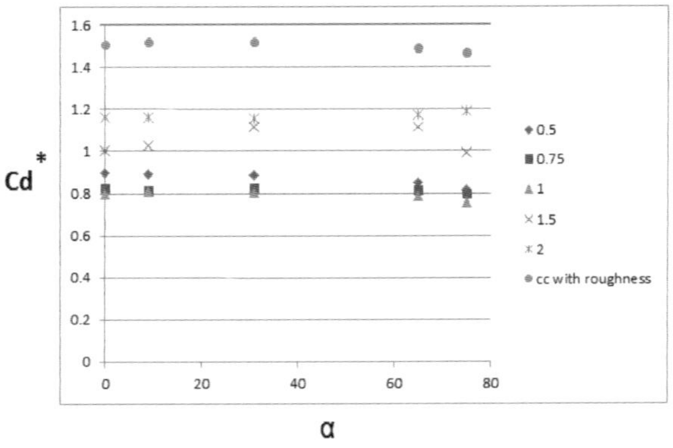

Figure.3: Variation of non-dimensional drag coefficient with roughness location

Fig.3 gives the non-dimensional drag coefficients for the study configuration (d) mentioned earlier (with both roughness and splitter plate). As could be seen, compared to the case with roughness alone, all the cases with splitter plate show a reduction in drag values. As G/D decreases, drag reduces in general. Quite interestingly, as G/D is further reduced from 1.0, drag slightly increases indicating that, there is an optimum G/D where drag is a minimum. This trend reversal is particularly evident at $\alpha = 0$ (stagnation point). At a given G/D value, change of strip locations does not bring any notable variation in the drag values. But, with roughness alone (no splitter plate), as Fig.3 shows, drag gets multiplied by a factor of about 1.5 compared to a smooth cylinder. Furthermore, Fig.3 also reveals that, a magnificent reduction of about 50% could be achieved by employing both roughness and splitter plate.

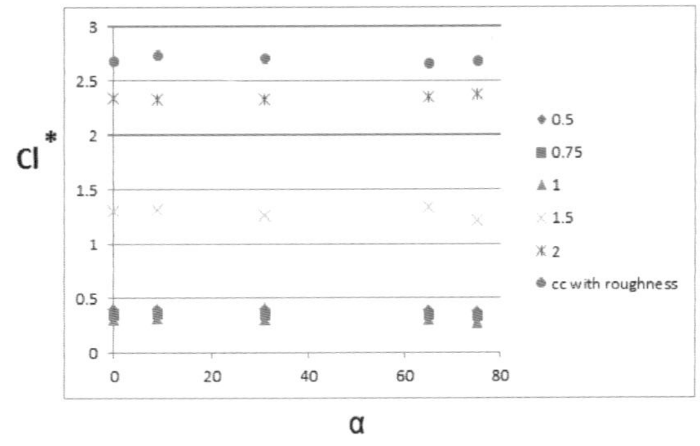

Figure.4: Variation of non-dimensional lift coefficient with roughness location

Non-dimensional lift coefficient (Fig.4) also shows a similar trend to that of the drag when both roughness and splitter plate are introduced. It could be seen that, with roughness (no splitter plate), lift is magnified by about 2.7 times that of smooth cylinder. Quite interestingly, as these results show, with the introduction of splitter plate along with roughness, lift coefficient drops to a substantially lower value equal to the value obtained in splitter plate alone configuration, i.e., 25% that of the smooth cylinder. Similarly, hike in drag coefficient obtained by roughness application (1.5 times that of smooth cylinder; Fig.3) is neutralised by the splitter plate resulting in a very low drag value equal (75% that of smooth cylinder). That is, selectively applied roughness tends to enhance both drag and lift whereas splitter tends to subdue both of them thus inducing mutually opposite effects. Therefore, it is thought that, by suitably employing roughness and splitter plate, the aerodynamic/hydrodynamic characteristics could be suitably manipulated or controlled. This meets the objective of the present study.

With splitter plate, lift drops down to lower values at lower G/D ratios; the drop is almost 75% (compared to smooth cylinder) at G/D=1.0. Similar to drag, a trend reversal occurs for lift also whereby reduction of G/D below 1.0 slightly increases the lift at all roughness locations tested. It is pointed out that, the Reynolds number of the current study (=17000) falls in the TrSL 2 regime where fluctuating lift shows notable magnification (Zdravkovich [15]) but it is lower than the magnification of the same compared to the TrSL 3 regime.

It is quite interesting to note that, at high Reynolds numbers (order of magnitude the same as that in the present study), selectively applied surface roughness induces high amplitude galloping oscillations in a flexibly mounted circular cylinder in water flow at majority of the circumferential locations considered by Chang et al [10] except at 64 deg (near the separation point). This seems to sensibly comply with the substantial hike in the lift coefficient obtained at almost all circumferential locations in the present study.

The observed features in the trends of drag, lift and Strouhal number characteristics of the test circular cylinder could be attributed to the structure of separated shear layers and its interaction with the roughness element and splitter plate. This has direct consequence on the wake dynamics and is reflected in Figs 2-4. However, to pinpoint the exact reasons, further experimental or numerical studies are required which is planned for the future.

5. Conclusions:

From the results obtained, the following conclusions are drawn:

(1) Roughness strip application is found to significantly alter the lift and drag forces acting on a circular cylinder. Lift coefficient is found to magnify about 2.7 times and the drag coefficient by about 1.5 times compared to a smooth cylinder. But, the circumferential location of the strip is found to have only a negligibly small influence on these parameters.

(2) With splitter plate, lift and drag coefficients reduce to lower values particularly at closer splitter plate positions (smaller G/D ratios). At G/D=1.0, lift and drag coefficients become the lowest.

(3) Present study reveals that surface roughness and splitter plate could be suitably employed to control the wake characteristics of a circular cylinder.

References

[1] Bishop R E D, Hassan A Y 1964 *Proceedings of the Royal Society of London Series* **A 277** 32–50
[2] Niemann H J, Holscher N 1990 *J.Wind Engg & Industrial Aerodynamics* **33** 197-209
[3] Gerrard J H 1966 *J. Fluid Mechanics* **25** 401-413
[4] Achenbach E 1971 *J. Fluid Mechanics* **46** 321–335
[5] Xu Y, Dalton C and Owen J C 2001 *J. Fluids and Structures* **15** 617-628
[6] Noberg C 2003 *J. Fluids and Structures* **17** 57-96
[7] Li Chen W, Gao L D, Yuan,Y W, Hui Li 2015 *Experiments in Fluids* **56** 1-15
[8] Dehkordi B G, Jafari H H 2010 *ASME J.Fluids Engg* **132** 1 - 4

IOP Conf. Series: Materials Science and Engineering **234** (2017) 012007 doi:10.1088/1757-899X/234/1/012007

[9] Assi G R S, Bearman P W 2015 *J.Fluids & Structures* **54** 263-280

[10] Chang C C, Ajithkumar R and Bernitsas M M 2011 *Ocean Engg* **38** 1713-1732

[11] Park H, Ajithkumar R and Bernitsas M M 2013 *Ocean Engg* **72** 403-415

[12] Park H, Ajithkumar R and Bernitsas M M 2016 *Ocean Engg* **111** 218-233

[13] Ozono S 1999 *Phy. of fluids* **11** 10

[14] Khan W A, Culham J R and Yovanovich M M 2003 *41st aerospace sciences meeting and exhibit*, 6-9 January, Reno, Nevada.

[15] Zdravkovich M M 1990 *J.Wind Engg & Industrial Aerodynamics* **33** 53-62

IOP Conf. Series: Materials Science and Engineering **234** (2017) 012008 doi:10.1088/1757-899X/234/1/012008

Detection of Cracks in a Cantilever Beam Using Signal Processing and Strain Energy Based Model

P Mehta[1], A Kureshi[2], S Lad[3], N Patel[4*] and D Sharma[5]

[1,2,3,4]Mechanical Department, School of Technology, Pandit Deendayal Petroleum University, Gandhinagar, Gujarat, India, 382007
[5]Mechanical Engineering Department, Faculty of Technology and Engineering, The M. S. University of Baroda, Gujarat, India

* nirav.npp@gmail.com

Abstract. Structure health monitoring is one of the most important aspects in an industry, as structures should work safely during their service life. Cracks are the most common damage that initiates a breakdown phase and hence timely and accurate detection of these cracks is imperative. In this article, a vibration based non-destructive technique is presented to detect one or multiple edge cracks in beam like structures. This model is based on variation in mode shapes and natural frequencies that provide accuracy in results as well as ease in practical applications. The crack location is identified using mode shapes of damaged beam wherein an appropriate signal processing technique is implemented by using which the noise in the signal can be reduced. Along with this, the crack severity is also determined using a strain energy based mathematical model. The model presented in this study is capable of detecting an arbitrary number of cracks in cantilever or simply supported configuration. The results obtained using the proposed method is also validated by considering few case studies.

1. Introduction

For any structure to work safely during its service life, it is necessary to assess its health on a regular basis. Hence, the continuous monitoring of the variation in its static or dynamic behavior is imperative. Mostly, these changes are directly influenced by the reduction of stiffness of the structure and in most cases, this reduction is caused by crack like defects. The presence of cracks in structure may affect the dynamic behavior like change in mode shapes, natural frequency, stiffness and damping ratio. The vibration based non-destructive technique (NDT) works upon this measured response of mode shapes and frequency. The use of vibration based technique has been explored by many researchers [1–6] for the detection of single and multiple cracks in various structures. Various articles implemented different approaches to model the cracks in beams. Some of the authors [7–11] replaced the crack by rotational spring while others [12–14] replaced by reduction in strain energy caused by a crack. Each method has its own specific advantages, although it is very difficult to satisfy all demands due to different types of changes in structures with various loading conditions. The spring based model has been applied to various types of beams like, stepped cantilever beam [7], geometrically segmented beams [8], simply supported beams [9], cantilever beam [10], multi-span beams [11] etc. The strain energy based model [12–14] has been used for crack detection in a cantilever beam of variable cross section. The basic idea of this model is to account for the energy absorbed by the crack and hence the crack can be substituted by any energy absorbing media. The strain energy based model works on the energy parameter itself and hence it is better suited to model cracked structures. Now, the variation in frequency and mode shapes are used in the crack detection. This variation can be seen as a discontinuity. The sensitivity of wavelet transform to discontinuities caused by the shift of deflection function and its capability to increase the resolution of the singularity signal has led to considerable developments in utilizing wavelet-based damage detection methodologies. Wang and Deng [14] were the first to address this possibility of determining the crack features of location and depth of damage in beams. The wavelet transform for crack detection has been further utilized by many researchers [15–19].

The aim of this paper is to establish an accurate method to detect and locate cracks in cantilever beam using wavelet transform. The methodology to detect crack severity is also presented along with the location of crack by using a strain energy based model. The model can be used to detect one or more cracks in a cantilever beam. The model is also validated by considering few case studies.

2. Methodology to identify Crack

Methods that detect damages directly from mode shapes are not accurate since inherent noise present in the signal can lead to incorrect detection and determination of damages also surrounding environment at times can introduce frequency changes in the structure greater than that introduced by the damage. Hence, suitable signal processing needs to be done on the obtained mode shape to achieve accuracy. Discontinuities are introduced in the mode shape of a damaged beam due to the reduction in strain energy of the beam. The wavelet transform is a signal processing technique which is accurate in identifying the discontinuities in the signal and hence can be effectively used onto the mode shapes of a damaged beam in order to locate damages. The wavelet transform signal consists of noticeable peaks wherein the signal has a discontinuity. A crack is described using two variables: location x_n and depth a_n for n^{th} crack, respectively as shown in figure 1. In the present study, the case of multiple cracks is considered that leads to the identification of any number of cracks.

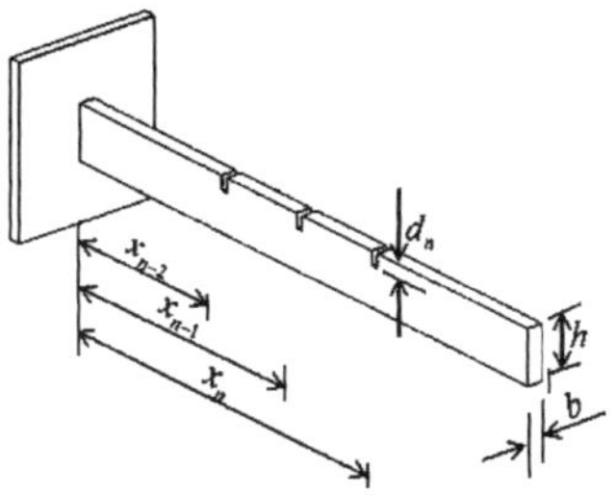

Figure 1. Beam Schematic

A mathematical model that describes the relationship between natural frequency variation and damage characteristic is presented in this article. The natural frequency of a beam can be derived considering the transverse vibration Euler-Bernoulli beam theory (beam is of constant cross section and rigidity). After using the displacement function as a method of separation of variables and putting it to the differential equation of a beam (Euler-Bernoulli beam theory), the following equation can be derived:

$$EI\Phi^{iv}(x) - \rho A\omega^2\Phi(x) = 0. \tag{1}$$

The solution of equation (1) is given as:

$$\Phi_i(x) = C_1\sin(\lambda x) + C_2 cos(\lambda x) + C_3\sinh(\lambda x) + C_4 cosh(\lambda x), \tag{2}$$

where, $\lambda^4 = \frac{\rho A\omega^2}{EI}$, ($A$ is cross sectional area, I is moment of inertia, Φ_i is the displacement function for i^{th} harmonic and $C_{1,2,3,4}$ are constants).

Applying the boundary conditions of a cantilever beam to the equation (2), the following equation of mode shape can be derived:

$$\Phi_i(x) = \cos(\lambda_i x) - \cosh(\lambda_i x) - \frac{\cos(\lambda_i L) + \cosh(\lambda_i L)}{\sin(\lambda_i L) + \sinh(\lambda_i L)}\left(\sin(\lambda_i x) - \sinh(\lambda_i x)\right). \tag{3}$$

The mode shapes presented in equation (3) can directly influence the strain energy. The strain energy (W_i) and natural frequency ($f_{h,i}$) of healthy beam can be written in terms of the mode shapes as below [12]:

$$W_i = \frac{1}{2}\int_0^L EI\left[\Phi''_i(x_1)\right]^2 dx;\ f_{h,i} = \int_0^L EI\left[\Phi''_i(x_1)\right]^2 dx, \tag{4-5}$$

where, W_i and $f_{h,i}$ are the strain energy and natural frequency of the healthy beam in the ith harmonic respectively. It can be observed from equations (4) and (5) that strain energy and frequency both are in direct proportional to the square of the second derivative of the mode shape. Similarly, in the case of the damaged beam, the strain energy from the section is reduced because of damage which results in the further reduction of natural frequency. Strain energy stored in a small section dx can be represented by equation (6):

$$dW_i(x_n) = \frac{1}{2}EI\left[\Phi''_i(x_n)\right]^2 dx. \tag{6}$$

It can also be interpreted from the equation (6) that a certain vibration energy change depends on the position of the damage on the beam i.e. the longitudinal location x_n, being influenced by the second derivative of the mode shape at the location while the depth of the crack only amplifies the response. As frequency is proportional to strain energy, change in frequency would be proportional to change in strain energy. Δf_i is the difference between $f_{h,i}$ (frequency of healthy beam) and $f_{c,i}$ (frequency of a cracked beam) in i^{th} mode of vibration ($\Delta f_i = f_{h,i} - f_{c,i}$). In the case of a cantilever beam having more than one crack, the total resultant change in modal strain energy can be considered as the summation of the individual change in modal strain energy caused by every single crack [20]:

$$\frac{\Delta f_i}{f_{h,i}} = \sum_{j=1}^{n} p\left(\overline{d}_i\right)\left[\Phi''_i(x_j)\right]^2. \tag{7}$$

$p(\overline{d}_i)$ reflects the influence of the depth of the crack while the term $\Phi_i''(x_i)$ reflects the influence of the location of the crack. This relation is valid for different types of configuration as well as dimensions of the beam considering suitable beam theories. The value of $p\left(\overline{d}_i\right)$ can be written as below [5]:

$$p\left(\overline{d}_i\right) = 6\pi(1-\upsilon^2)\begin{pmatrix} 0.6272\left(\overline{d}_i\right)^2 - 1.04533\left(\overline{d}_i\right)^3 + 4.5948\left(\overline{d}_i\right)^4 - 9.9736\left(\overline{d}_i\right)^5 + 20.2948 \\ \left(\overline{d}_i\right)^6 - 33.0341\left(\overline{d}_i\right)^7 + 47.163\left(\overline{d}_i\right)^8 - 40.7556\left(\overline{d}_i\right)^9 + 19.6\left(\overline{d}_i\right)^{10} \end{pmatrix}, \tag{8}$$

where, $\overline{d}_i = d_i/h$ is a dimensionless damage depth and the functions $p\left(\overline{d}_i\right)$ is stress intensity factor. Using the values of equation (8), one can write equation (7) for different mode shapes:

$$\left\{\frac{\Delta f_1}{f_{h,1}}, \frac{\Delta f_2}{f_{h,2}}, \ldots \frac{\Delta f_n}{f_{h,n}} = \sum_{j=1}^{n} p\left(\overline{d}_j\right)\left[\Phi''_1(x_j)\right]^2, \sum_{j=1}^{n} p\left(\overline{d}_j\right)\left[\Phi''_2(x_j)\right]^2 \ldots \sum_{j=1}^{n} p\left(\overline{d}_j\right)\left[\Phi''_n(x_j)\right]^2.\right. \tag{9}$$

Solving the equation (9), one can obtain the values for $\{d_1, d_2 \ldots d_n\}$. Overall process to detect location and depth can be summarized in figure 2.

3. Validation Study

The methodology to detect one or multiple cracks in a cantilever beam is presented in this article. The model can be validated by considering a beam with various configurations of cracks. The cantilever beam with Young's Modulus (E) = 70.09 GPa, density (ρ) = 2700 kg/m^3, length (L) = 0.24 m, height (h) = 0.0191 m, width (b) = 0.0064 m and moment of Inertia (I) = 3716 mm^4 is considered in present study. The mode shapes of the damaged beam can be obtained by experimentation or by using finite element analysis based commercial software package. In the present study, a cantilever beam has been used with two edge cracks at normalized location 0.25 and 0.5 with normalized depths 0.35 and 0.45 respectively. The finite element analysis model is presented in figure 3.

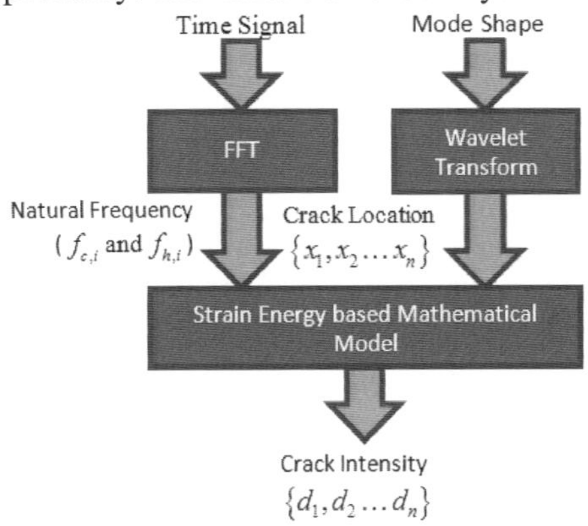

Figure 2. Flow chart of crack detection method

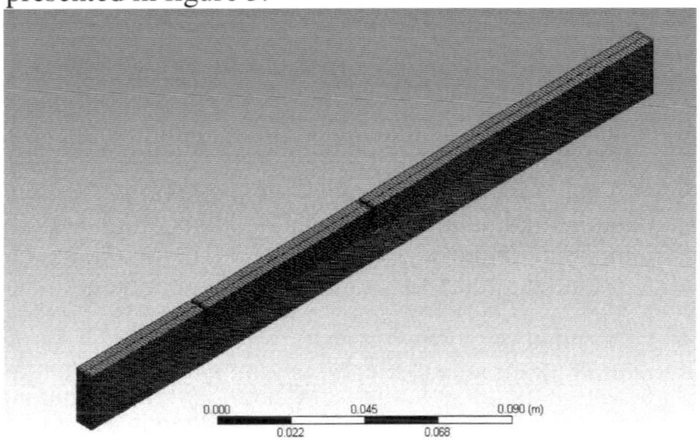

Figure 3. FEA model of a cantilever beam

Applying proper boundary conditions in FEA model (figure 3), the mode shape or displacement can be obtained for any type of damaged beam. Such displacement plot is presented in figure 4. Figure 5 shows the wavelet transform of the second mode shape of the damaged beam. It can be seen clearly that the wavelet transform picked up a disturbance in the mode shape at the normalized location of the crack: x_1 = 0.24 and x_2 = 0.49, which is then compared to the actual location. Similarly, the results are also validated for different locations of crack (table 1). It is evident that when the location of the crack is near the fixed end, the error in locating the crack increases since near the fixed end the displacement of the beam would be less and subsequently, the discontinuity in the mode shape would also be less. Additionally, the severity of the crack can also be identified in the present model using equation (9). The different cases of crack depth (at the same location as table 1) have been considered. The detected location from table 1 and the natural frequency are inserted in the mathematical model to obtain the crack severity. The comparison is listed in table 2.

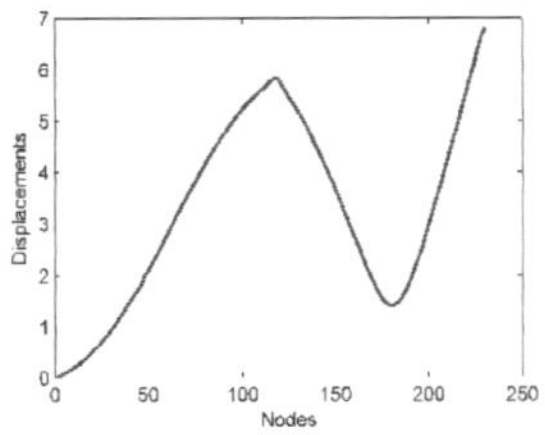

Figure 4. Second Mode Shape of the damaged beam

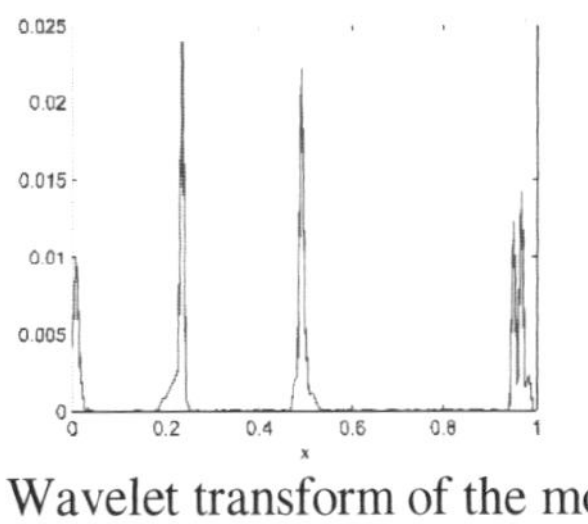

Figure 5. Wavelet transform of the mode shape

Table 1. Crack Location

Sr No	Actual Normalized Location		Detected Normalized Location		Error	
	Crack 1	Crack 2	Crack 1	Crack 2	Crack 1	Crack 2
1	0.25	0.5	0.24	0.49	-4	-2

2	0.2	0.5	0.2	0.49	0	-2
3	0.25	0.5	0.24	0.48	-4	-4
4	0.1	0.5	0.11	0.5	9	0
5	0.25	0.5	0.24	0.49	-4	-2
6	0.15	0.5	0.15	0.51	0	2

Table 2. Crack Severity

Sr No	Natural Frequency		Normalized Actual Crack Depth		Normalized Detected Crack Depth		Error	
	Mode 1	Mode 2	Crack 1	Crack 2	Crack 1	Crack 2	Crack 1	Crack 2
1	237.7	1415.4	0.35	0.45	0.35	0.43	0.89	-3.89
2	232.5	1590.7	0.45	0.25	0.46	0.26	3.13	3.76
3	208.02	1309.9	0.45	0.45	0.44	0.42	-2.23	-6.67
4	231.9	1636.9	0.46	0.15	0.44	0.14	-3.04	-3.46
5	240.29	1393.8	0.3	0.45	0.3	0.41	0	-8
6	233.07	1521.1	0.43	0.35	0.43	0.33	-0.46	-4.85

The mean error (table 2) is 3.8% which is quite accurate for such Non-Destructive Technique. It is evident that the measurements were consistent up to $\bar{d} = 0.5$. The vibration coupling (at larger depths) resulted in a crowded spectrum and a complex time signal that influence the accuracy of natural frequency. Moreover at depths smaller than $\bar{d} = 0.1$, the difference from the natural frequency of the un-cracked beam was not measurable. Therefore, the results will be limited to $0.1 \le \bar{d} \le 0.6$ for present case [5].

4. Conclusion

A quick, accurate and reliable vibration based non-destructive technique to detect cracks is presented in this study. The method can be applied to identify any number of cracks in a cantilever beam. The methodology can also be extended to the various configuration of beams with a change in mode shapes. The method presented in this study is also validated by considering various cases of cracks locations and crack depths. The method can be used by any designer who wants to detect cracks in beam like structures.

References

[1] Adams RD, Cawley P, Pye CJ, Stone BJ. A Finite Element Approach to Plate Vibration Problems. *J Mech Eng Sci.* 1965;7(1):5.
[2] Cawley P. The Location of Defects in Structures From Measurements of Natural Frequencies. *J strain Anal.* 1978;14(2):49–57.
[3] Dimarogonas AD, Papadopoulos CA. Vibration of cracked shafts in bending. *J Sound Vib.* 1983;91(4):583–93.
[4] Shen MH, Pierre C. Free Vibrations of beams with single edge crack. *Jounral Sound Vib.* 1994;170(2):237–59.
[5] Chondros TG, Dimarogonas A. D, Yao J. A Continuous Cracked Beam Vibration Theory. *J Sound Vib* [Internet]. 1998;215:17–34. Available from: http:// www.sciencedirect.com /science/ article/ pii /S0022460X98916406
[6] Dimarogonas AD. Vibration of cracked structures: A state of the art review. *Eng Fract Mech.* 1996;55(5):831–57.
[7] Nandwana BP, Maiti SK. Detection of the Location and Size of a Crack in Stepped Cantilever Beams Based on Measurements of Natural Frequencies. *J Sound Vib* [Internet]. 1997;203(3):435–46. Available from: http://www.sciencedirect.com/science/article/pii/S0022460X96908561
[8] Chaudhari TD, Maiti SK. A study of vibration of geometrically segmented beams with and

without crack. *Int J Solids Struct* [Internet]. 2000;37(5):761–79. Available from: http://www.sciencedirect.com/science/article/pii/S0020768399000542

[9] Patil DP, Maiti SK. Detection of multiple cracks using frequency measurements. *Eng Fract Mech*. 2003;70(12):1553–72.

[10] Patil DP, Maiti SK. Experimental verification of a method of detection of multiple cracks in beams based on frequency measurements. *J Sound Vib*. 2005;281(1–2):439–51.

[11] Sharma DS, Mungla MJ, Barad KH. Vibration-based non-destructive technique to detect crack in multi-span beam. *Nondestruct Test Eval* [Internet]. 2015;30(4):291–311. Available from: http://www.tandfonline.com/doi/full/10.1080/10589759.2015.1029475

[12] Gillich G-R, Praisach Z-I. Modal identification and damage detection in beam-like structures using the power spectrum and time–frequency analysis. *Signal Processing* [Internet]. 2014;96:29–44. Available from: http://www.sciencedirect.com/science/article/pii/S0165168413001722

[13] Yang XF, Swamidas ASJ, Seshadri R. Crack Identification in Vibrating Beams Using the Energy Method. *J Sound Vib* [Internet]. 2001;244(2):339–57. Available from: http://linkinghub.elsevier.com/retrieve/pii/S0022460X00934989

[14] Wang Q, Deng XM. Damage detection with spatial wavelets. *Int J Solids Struct*. 1999;36:3443–68.

[15] Douka E, Loutridis S, Trochidis A. Crack identification in beams using wavelet analysis. *Int J Solids Struct*. 2003;40(13–14):3557–69.

[16] Chang CC, Chen LW. Detection of the location and size of cracks in the multiple cracked beam by spatial wavelet based approach. *Mech Syst Signal Process*. 2005;19(1):139–55.

[17] Ovanesova A V., Suárez LE. Applications of wavelet transforms to damage detection in frame structures. *Eng Struct*. 2004;26(1):39–49.

[18] Khatam H, Golafshani a. a., Beheshti-Aval SB, Noori M. Harmonic Class Loading for Damage Identification in Beams Using Wavelet Analysis. *Struct Heal Monit*. 2007;6(1):67–80.

[19] Xiang J, Liang M. Wavelet-Based Detection of Beam Cracks Using Modal Shape and Frequency Measurements. *Comput Civ Infrastruct Eng*. 2012;27(6):439–54.

[20] Kisa M. Free vibration analysis of a cantilever composite beam with multiple cracks. *Compos Sci Technol*. 2004;64(9):1391–402.

IOP Conf. Series: Materials Science and Engineering **234** (2017) 012009 doi:10.1088/1757-899X/234/1/012009

Development of Double Layer Microwave Absorber Using Genetic Algorithm

Abhishek Kumar[1], Samarjit Singh[1] and Dharmendra Singh[2]

[1] Department of Applied Mechanics, Motilal Nehru National Institute of technology Allahabad, Allahabad-211004, India
[2] Department of Electronics and Communication, Indian Institute of Technology Roorkee, Roorkee-247667, India

abhishek@mnnit.ac.in

Abstract. In this paper, an efficient two-layer microwave absorber at X-band is designed, optimized and implemented using the available materials with frequency dependent complex permittivity and complex permeability values as material database. The present work is focused on the design of a two- layer microwave absorber with good microwave absorption properties combined with broadband features at X-band. The optimization of various parameters such as materials, their sequence and thickness for obtaining better microwave absorption characteristics at X-band has been realized using Genetic Algorithm (GA). The optimized results were used to design a two-layer microwave absorber and experimentally tested using Attenuation Testing Device (ATD). Further verification of the experimentally obtained absorption results were simulated in High Frequency Structure Simulator (HFSS). The ATD result show that the maximum Reflection Loss (RL) for two-layer microwave absorber was -21.98 dB with 2.77 GHz bandwidth (corresponding to -10 dB) at 11.06 GHz for a total coating thickness of 1.5 mm.

1. Introduction

In recent years, the use of microwave signals has been very significant due to the mammoth increase in the use of electronic devices especially in areas such as communication systems, radar and stealth technology used in military spheres [1]. This technological advancement has led to the advent of microwave absorbing materials. Due to the rapid growth in the use of microwave absorbers, there is a continuous need to improve their performance in order to attain better microwave absorption properties. In response to the need, an effective microwave absorber requires more and more attention towards improving the performance of the already existing materials by tailoring them. Different types of materials have been used for microwave absorption, which include magnetic and dielectric (non-magnetic) absorbers. The microwave absorption characteristic of these materials is dependent on their complex permittivity and permeability values for the concerned frequency range. One of the major concerns in developing a microwave absorber is to obtain broadband characteristics. The need to design microwave absorbers with broadband characteristics has led to the concept of multi-layering. In multi-layering, thin microwave absorbing coatings are stacked one over another to gain broadband characteristics. For designing multi-layer microwave coatings, the number of layers contained in the coating, material choice for each layer and their thickness need to be optimized from a predefined set of available materials, over a specified frequency range [2]. The material choice for each layer is quite critical due to impedance matching. An impedance match promotes the penetration of the incident microwave radiation into the material and thereafter absorption of the microwave energy takes place which attenuates the incident microwave radiations [3]. Understanding the various phenomena that act on multi-layer absorbers is complex. The complexity in understanding the mechanisms arises from the diverse impedances involved in the formation of multi-layered structures which results in various ways the incident microwave can interacts which each layer. To take into consideration these complexities, computer simulation is necessary for the resolution of complex algorithms [4].

In this work, design and analysis of two-layer microwave absorber has been carried out. The synthesized materials with co mplex permittivity and complex permeability values were used as the material database for evaluation of the microwave absorption characteristics in the X-band. The optimization of various parameters such as materials, their sequence and thickness of each layer has been successfully carried out using Genetic Algorithm (GA). The optimized values were simulated using High Frequency Structure Simulator (HFSS) and the experimental measurements were done using Attenuation Testing Device (ATD).

2. Theoretical background of multi-layering

The multi-layering approach is very critical in enhancing the broadband characteristics of the microwave absorber. It is also an important tool to reduce the overall coating thickness of the absorber. The multi-layering technique increases the number of boundaries and interfaces [5]. These boundaries and interfaces acts as site for multiple internal reflections thereby increasing the probability of phase cancellation and hence improving the microwave absorption characteristics. Besides this, interfacial polarization occurs due to the entrapment of charge carriers at the interfaces of the heterogeneous media in multi-layering. The charge transfer between the various phases present in the multi-layered material may give rise to the interfacial polarization and the associated relaxations which contribute to the enhanced microwave absorption performance of the heterogeneous multi-layered system. The phenomenon of multi-layering has been clearly depicted in Figure 1.

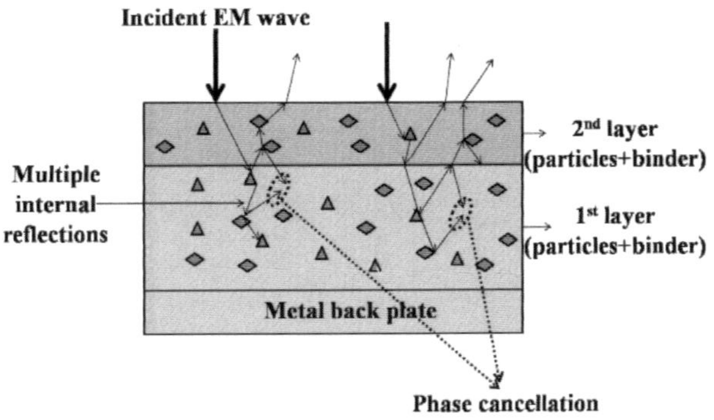

Figure 1. A cartoon depicting multi-layering phenomenon for a double layer microwave absorber.

2.1. Mathematical formulation for multi-layering approach

An infinite conducting plate coated with multilayer microwave absorber having layers of different coating thickness with different complex permittivity and complex permeability is shown in Figure 2. Let an electromagnetic (EM) wave is incident perpendicular to the multi-layer absorber and propagates in the material. Coating parameters t_i, ε_i, μ_i and z_i represents the thickness, complex permittivity, complex permeability and intrinsic impedance, respectively of the ith layer for multi-layer absorber. The first layer is in contact with a perfectly conducting metal plate and the mth layer is terminated to free space. Let ε_o, μ_o and z_o are the permittivity, permeability and impedance of free space respectively.

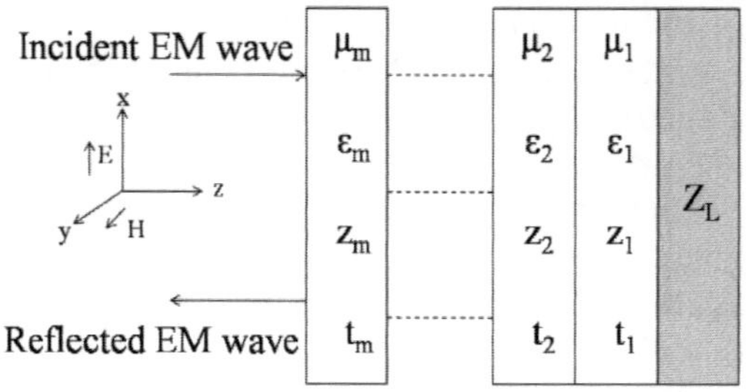

Figure 2. Schematic diagram of multi-layer microwave absorber.

The RL for multi-layering is calculated using the transmission line theory and is mathematically formulated as [6]:

$$\Gamma = \frac{Z_{in,m} - Z_0}{Z_{in,m} + Z_0} \tag{1}$$

The reflection coefficient for two-layer absorber, where layer 1 is backed with perfect conductor and layer 2 is in contact with free space, is written as:

$$\Gamma = \frac{Z_{in,2} - Z_0}{Z_{in,2} + Z_0} = \frac{\dfrac{\sqrt{\mu_1/\varepsilon_1}\tanh(\gamma_1 t_1) + \sqrt{\mu_2/\varepsilon_2}\tanh(\gamma_2 t_2)}{1 + \sqrt{\mu_1\varepsilon_2/\mu_2\varepsilon_1}\tanh(\gamma_1 t_1)\tanh(\gamma_2 t_2)} - 1}{\dfrac{\sqrt{\mu_1/\varepsilon_1}\tanh(\gamma_1 t_1) + \sqrt{\mu_2/\varepsilon_2}\tanh(\gamma_2 t_2)}{1 + \sqrt{\mu_1\varepsilon_2/\mu_2\varepsilon_1}\tanh(\gamma_1 t_1)\tanh(\gamma_2 t_2)} + 1} \tag{2}$$

$$Z_{in,2} = (Z_{in,1} + Z_2 \tanh(\gamma_2 t_2))/((Z_{in,1}/Z_2)\tanh(\gamma_2 t_2) + 1) \tag{3}$$

$$Z_2 = Z_0\sqrt{\frac{\mu_2}{\varepsilon_2}} \tag{4}$$

where, $Z_{in,2}$ input impedance for two-layer absorber at air/absorber interface and Z_2 impedance of second layer for two-layer absorber and

$$\gamma = j\frac{2\pi f}{c}\sqrt{\mu\varepsilon} \tag{5}$$

The reflection loss (RL) of the incident EM wave normal to the multilayer absorber in dB is expressed as [7]:

$$\text{Reflection Loss (dB)} = -20\log_{10}|\Gamma| \tag{6}$$

2.2. Parameter optimization using GA

The theoretical background and mathematical formulations gives us an indication that multi-layering is a very complex process which involves selection of suitable materials, their sequence and thickness of each layer for better microwave absorption characteristics. These parameters can be easily optimized with the help of Genetic Algorithm (GA). GA is a powerful optimization tool for problems having a large number of variables and has been extensively used by researchers for solving multi-layering problems. In the present work, "gamultiobj - multiobjective optimization using Genetic Algorithm" has been used to optimize different variables for designing a two-layer microwave absorber, as it implements the genetic algorithm at the command line to minimize a multi-component objective function. The flowchart for GA optimization is shown in Figure 3(a) and the material database used has been shown in Table 1. The parameters used in 'gatool' for optimization is given in

Table 2. The following fitness function (F) has been maximized to optimize the different parameters i.e., to get maximum reflection loss (RL) values using GA

$$F = |RL_{computed} - RL_{observed}|^2 \quad (7)$$

where, RLobserved is the required value fixed by the user and RLcomputed is calculated from the material database (Table 1) using Eqs. 2 to 6.

Table 1. Material database used for optimization of two-layer absorber [8]

S No	Material	Code	S No	Material	Code	S No	Material	Code
1	$BaFe_{12}O_{19}$ (2 h)	P_1	11	SiC (1 h milled)	S_3	21	Al_2O_3	M_{10}
2	$BaFe_{12}O_{19}$ (4 h)	P_2	12	SiC (2 h milled)	S_4	22	Al_2O_3 + 10 wt.% Al	M_{11}
3	$BaFe_{12}O_{19}$ (8 h)	P_3	13	SiC (3 h milled)	S_5	23	Al_2O_3 + 10 wt.% Ni	M_{12}
4	$BaCo_{0.2}Fe_{11.8}O_{19}$	Q_2	14	SiC (6 h milled)	S_6	24	SiC + 10 wt.% Al	M_2
5	$BaCo_{0.4}Fe_{11.6}O_{19}$	Q_3	15	SiC (10 h milled)	S_7	25	SiC + 10 wt.% Co	M_3
6	$BaCo_{0.6}Fe_{11.4}O_{19}$	Q_4	16	SiC (20 h milled)	S_8	26	SiC + 10 wt.% Cr	M_4
7	$BaCo_{0.8}Fe_{11.2}O_{19}$	Q_5	17	SiC (micro wire)	R_3	27	SiC + 10 wt.% Mn	M_5
8	$BaCo_{1.0}Fe_{11.0}O_{19}$	Q_6	18	TiO_2	M_{40}	28	SiC + 10 wt.% Ni	M_6
9	SiC (200-325 mesh)	S_1	19	TiO_2 + 10 wt.% Al	M_{41}	29	SiC + 10 wt.% Ti	M_7
10	SiC (325-400 mesh)	S_2	20	TiO_2 + 10 wt.% Ni	M_{42}	30	SiC + 10 wt.% Zn	M_8

Table 2. Parameters used for optimization of a double layer absorber in gatool.

Entity	Selected parameters
Solver	Gamultiobj
Fitness function	@objfun
Number of variables	4
Lower bound	[0.5 0.5 0 0]
Upper bound	[2 2 29 29]
Population type	Double vector
Selection function	Tournament
Mutation function	Adaptive feasible
Cross over function	Two point
Direction	Both
Hybrid function	Fgoalattain

2.3. Modeling using High Frequency Structure Simulator (HFSS)

HFSS is a commercial finite element based solver for EM structures. In order to validate and analyze the GA optimized results, Ansys 'High Frequency Structure Simulator' (HFSS) tool has been used to compare the absorption with the experimental results in X-band (8.2-12.4 GHz). After optimization of various parameters for two-layer absorber, these GA optimized were used to design two-layer model backed with Al-metal in HFSS. The flowchart shown in Figure 3(b) depicts the various steps involved in the modeling of two-layer structure in HFSS.

2.4. Measurement of microwave absorption using Attenuation Testing Device (ATD)

The experimental measurements of the reflection loss of incident EM waves on the surface of the developed two-layer microwave absorber has been carried out using an ingeniously developed Attenuation Testing Device (ATD) for X-band similar to IEEE standard 1128-1998. This ATD setup consists of a pyramidal horn antenna with its aperture fitted with an extended wave guide terminated with a metallic variable short [9].

The reflected power (P_1) is noted without coating on reference conducting plate of size (72.4 x 96.4 mm^2). After that reflected power (P_2) with absorber coated reference conducting plate is calculated.

IOP Conf. Series: Materials Science and Engineering **234** (2017) 012009 doi:10.1088/1757-899X/234/1/012009

The difference in the two readings of power (P_1-P_2) gives power absorbed by the absorber. Thus, the absorption in dB can be written as:

$$\text{Absorption (dB)} = P_1 - P_2$$

(a)

Geometry of two-layer absorber is mathematically defined in terms of variables
↓
Calculate total impedance ($Z_{in,m}$) at outer layer-free space interface using Eq. 3 for two-layer absorber
↓
Calculate reflection loss using Eq. 6
↓
Optimize fitness function (F) defined in Eq. 7 using Genetic Algorithm
↓
Optimize parameters (Materials, their sequence and thickness)

(b)

Model design (which includes geometry and material selection)
↓
PML boundaries
↓
Master and slave boundary
↓
Excitation (Plane wave)
↓
Analysis (solution setup, Frequency sweep)

Figure 3. Flowchart depicting the steps involved in (a) optimization of parameters for a two-layer microwave absorber and (b) HFSS modelling.

3. Results and discussions

Two-layer absorber sheet has been fabricated as per the optimized results Table 3 and tested using ATD. The RL curve for two-layer microwave absorber is shown in Figure 4 and the corresponding RL data is listed in Table 4. The curve shows that the GA optimized materials M_{41} and M_6 have maximum RL of – 13.67 dB and -24.93 dB with absorption bandwidth of 1.01 and 2.10 GHz, respectively. However, when these materials are stacked to become a two-layer absorber with first layer of M_{41} (thickness 0.7 mm) and second layer of material M_6 (thickness 0.8 mm), the value of RL as well as its bandwidth increases (Table 4). The simulation RL results of HFSS model are represented by RL_{HFSS}. The experimental results represented by RL_{ATD} also verify the optimized RL values. A small variation in HFSS and ATD results has been observed which is due to fabrication error in coating of materials.

Table 3. GA optimized results for two layer absorber.

S No	Two layer absorber	GA optimized result
1	Ist layer material	M_{41}
2	Ist layer thickness	0.7 mm
3	IInd layer material	M_6
4	IInd layer thickness	0.8 mm

Table 4. Reflection loss results for two layer absorber.

Sample ID	Frequency Max. RL (GHz)	Thickness Max. RL (mm)	Max. RL (dB)	-10 dB absorption bandwidth (GHz)
M_{41}	10.13	1.3	-13.67	1.01
M_6	11.81	1.5	-24.93	2.10
RL_{HFSS}	10.72	0.7 + 0.8	-23.37	2.27
RL_{ATD}	11.06	0.7 + 0.8	-21.98	2.77

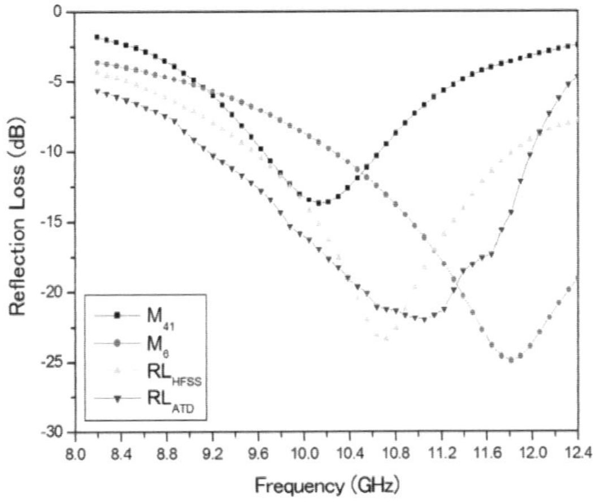

Figure 4. Reflection loss characteristic of two layer microwave absorber.

4. Conclusions

In the present work, a two-layer microwave absorber is designed using the materials from the material database and optimized using GA. The optimized GA result gives the materials, their sequence and thickness. The optimized results were used to develop a two layer microwave absorber and then experimentally tested using ATD in the X-band. Further verification of the tested results was simulated using HFSS tool. The results show that good RL characteristic with increase in broadband characteristic is observed for the two-layer microwave absorber. The attainment of coating thickness below 2 mm and RL values above -20 dB with increased bandwidth at X-band reveals excellent microwave absorption characteristic of the two-layer absorber.

References

[1] Folgueras L C and Rezende M C 2008 Multilayer radar absorbing material processing using polymeric nonwoven and conducting polymer Materials Research 11 245-249.

[2] Gargama H, Chaturvedi S K and Thakur A K 2012 Design and optimization of multi-layered electromagnetic field using a real coded genetic algorithm Progress in Electromagnetics Research B 39 241-266.

[3] El-Hakim H K, Mahmoud K R and Abdelaziz A A 2016 Design of compact double-layer microwave absorber for X-Ku bands using genetic algorithm Progress in Electromagnetics Research B 65 157-168.

[4] Parida R C, Singh D and Agarwal N K 2007 Implementation of multilayer ferrite radar absorbing coating with genetic algorithm for radar cross-section reduction at X-band Indian Journal of Radio and Space physics 36 145-152.

[5] Najim M, Smitha P, Agarwala V and Singh D 2015 Design of light weight multi-layered coating of zinc oxide-iron-graphite nano-composites for ultra-wide bandwidth microwave absorption J Mater Sci: Mater Electron 26 7367-7377.

[6] Perhi J and Cohen L S 1993 Design of broadband radar absorbing materials for large angle of incidences IEEE Transactions on Electromagnetic Compatibility 35 223-230.

[7] Kumar A, Agarwala V and Singh D 2014 Effect of milling on dielectric and microwave absorption properties of SiC based composites Ceramic International 40 1797-1806.

[8] Kumar A 2013 Design and development of coatings for RADAR absorbing materials at X-band Doctorate thesis Indian Institute of Technology Roorkee.

[9] Singh D, Kumar A, Meena S and Agarwala V 2012 Analysis of frequency selective surfaces for radar absorbing materials Progress in Electromagnetics Research B 38 297-314.

IOP Conf. Series: Materials Science and Engineering **234** (2017) 012010 doi:10.1088/1757-899X/234/1/012010

Study on Stationarity of Random Load Spectrum Based on the Special Road

Huawen Yan[1], Weigong Zhang[1,2] and Dong Wang[1]

[1]School of Instrument Science and Engineering, Southeast University, Nanjing 210096, China

[2]zhangwg@seu.edu.cn

Abstract. In the special road quality assessment method, there is a method using a wheel force sensor, the essence of this method is collecting the load spectrum of the car to reflect the quality of road. According to the definition of stochastic process, it is easy to find that the load spectrum is a stochastic process. However, the analysis method and application range of different random processes are very different, especially in engineering practice, which will directly affect the design and development of the experiment. Therefore, determining the type of a random process has important practical significance. Based on the analysis of the digital characteristics of road load spectrum, this paper determines that the road load spectrum in this experiment belongs to a stationary stochastic process, paving the way for the follow-up modeling and feature extraction of the special road.

1. Introduction

In probability theory and related fields, a stochastic or random process is a mathematical object usually defined as a collection of random variables. Stochastic processes are widely used as mathematical models of systems and phenomena that appear to vary in a random manner. They have applications in many technology and engineering fields such as image processing, signal processing, information theory and computer science. Historically, the random variables were associated with or indexed by a set of numbers, usually viewed as points in time, giving the interpretation of a stochastic process representing numerical values of some system randomly changing over time, such as growth of a bacterial population, but the system studied this paper is a stochastic process representing numerical values of road loads randomly changing over space. On a certain length of road, the loads applied to the car by the road are random variables based on space, and these loads in each replicate experiment are not the same, so that it is not one or a few random variables, but a family of infinite number of random variables [1]. Therefore, it can be determined that the load spectrum is a typical stochastic process according to the definition of a stochastic process. Based on their properties, stochastic processes can be divided into various categories, which include Markov processes, Wiener processes, Poisson processes, renewal processes and stationary processes. The study of stochastic processes requires mathematical knowledge and techniques from probability, calculus, linear algebra, and topology as well as branches of mathematical analysis such as real analysis, and Fourier analysis, while different stochastic process requires different mathematical knowledge and techniques, so in order to determine the kind of road load spectrum, this paper was divided into three steps. In short, first, put forward the framework of theoretical analysis; second, design an experiment according to the theory; third, make a validation of the theory using the data collected by the experiment.

2. Theoretical analysis

Despite limited distribution function family of stochastic processes can characterize the statistical properties of stochastic processes. However, it is often difficult and even impossible to determine the distribution function family of the stochastic process in practical application, so it is necessary to introduce the basic digital features of the random process as the numerical feature of the random variable is introduced. These digital features not only can characterize the important characteristics of the stochastic process, but also can facilitate the operation and the actual measurement. Therefore, four commonly used numerical features, which are mean function, variance function, correlation function, covariance function, are introduced to study the road load spectrum in this paper.

2.1 Mean Function and Variance Function

$$\mu(t) = E[X(t)] = \int_{-\infty}^{+\infty} xf(x;t)dx \tag{1}$$

$$\sigma^2 = D[X(t)] = E\{[X(t)-\mu(t)]^2\} = \int_{-\infty}^{+\infty}[x-\mu(t)]^2 f(x;t)dx \tag{2}$$

2.2 Correlation Function and Covariance Function

The mean and variance are digital features that characterize the statistical properties of the process at each isolated time, and the numerical features that can reflect the statistical characteristics of the process at different times is required, since new digital features are required to describe the stochastic process [2].

$$\rho(t1,t2) = E[X(t1)X(t2)] = \iint_{-\infty}^{+\infty} x1x2f(x1,x2;t1,t2)dx1dx2 \tag{3}$$

$$\gamma(t1,t2) = Cov(X(t1),X(t2)) = E\{[X(t1)-\mu(t1)][X(t2)-\mu(t2)]\} \tag{4}$$

From the statistical theory we can see that the covariance function and the variance can be expressed by the mean and the autocorrelation function, so the mean and autocorrelation functions are more important digital features. While, when the time parameter t is discrete, X(t) is the time series, and the load data studied in this paper belong to discrete time series.

2.3 Stationary Condition

The process $\{X_t, t \in Z\}$ is said to be stationary if it satisfies the following three properties [3]:

1) $E|X_t|^2 < \infty$ for all t∈Z,
2) $EX_t = m$ for all t∈Z,
3) $\gamma_X(t1,t2) = \gamma_X(t1+\tau, t2+\tau)$ for all t1, t2, τ∈Z.

An important consequence of property 3) is that the variance of the tth component X_t of the stationary process $\{X_t\}$ is a constant independent of t because $\gamma_X(t1,t2) = \gamma_X(t1-t2,0)$. Accordingly, defining covariance function and correlation function of stationary process to be a unary function for convenience

$$\gamma_X(h) \triangleq \gamma_X(h,0) = Cov(X_{t+h}, X_t), \text{ for all t, h } \in Z. \tag{5}$$

$$\rho(h) = \frac{\gamma_X(h)}{\gamma_X(0)} \quad |h| \le n, \text{ where n represents the number of samples}$$

3. Arrange the experiment

While a vehicle is on the road, because of the effect of the ground, some forces are applied to the wheel, namely the longitudinal force Fx, the lateral force Fy, the vertical force Fz[4], as shown in figure 1A. It is generally known that the role of force is mutual, so the wheel force is equivalent to the road load. The load can be collected by the WFT system, and the block diagram is shown in figure 1B. The hardware components of the system include elastomer, collection-module, encoder, transmission-module, central collection box, GPS module and laptop.

IOP Conf. Series: Materials Science and Engineering **234** (2017) 012010 doi:10.1088/1757-899X/234/1/012010

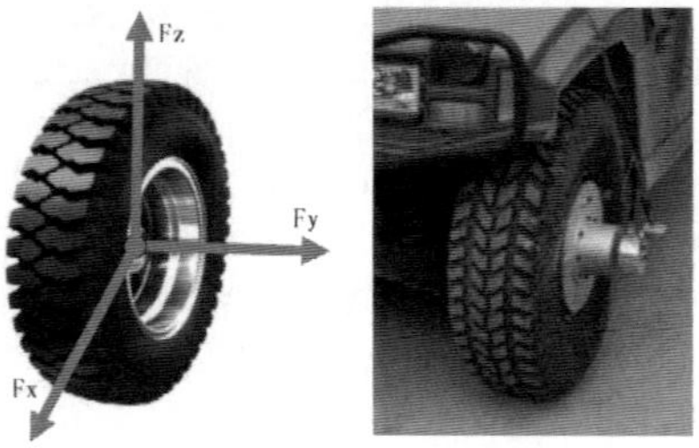

Figure 1.A: Forces diagram

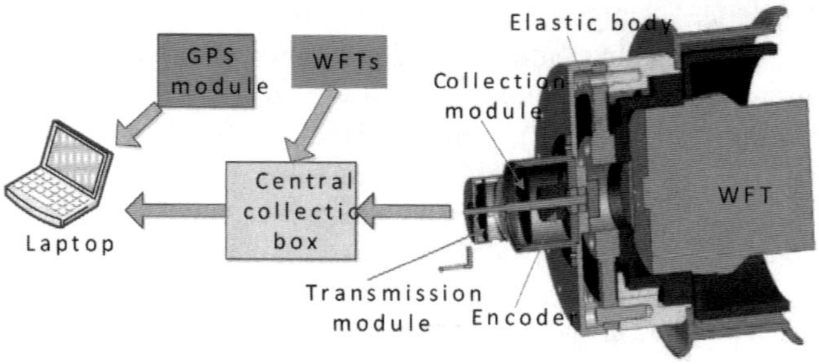

Figure 1.B: System block diagram

The WFT system is a vehicle-to-road coupling system, so the measured road load is related to multiple factors, which include road type, vehicle type, speed, tire pressure and so on. Road load of different road is not the same, which is the reason for setting various special roads. Different vehicles have different suspension types and weights, and their loads on the road are naturally different. In addition, the impact of speed and tire pressure on road loads is also evident. The vehicle used in this experiment is the military active off-road vehicle, which total weight is equal to curb weight plus weight of testers and equipment for the 3.7t, and the tire pressure was adjusted to 2.7Kg/cm^2. The sensors in the WFT system are mounted on the left and right wheels of the front axle. The speed of the vehicle is maintained at 40km/h, and GPS module is responsible for vehicle speed compensation. The special road used in the experiment was the cobblestone road, which is shown in figure. 2A, and a total of 3000 groups sample loads data were collected. One group of the original loads data collected by the left WFT are shown in figure. 2B to figure. 2D. Owning to the right WFT is similar to the left one, so the data curve of the left WFT is drawn alone for a brief description.

A : the cobblestone road

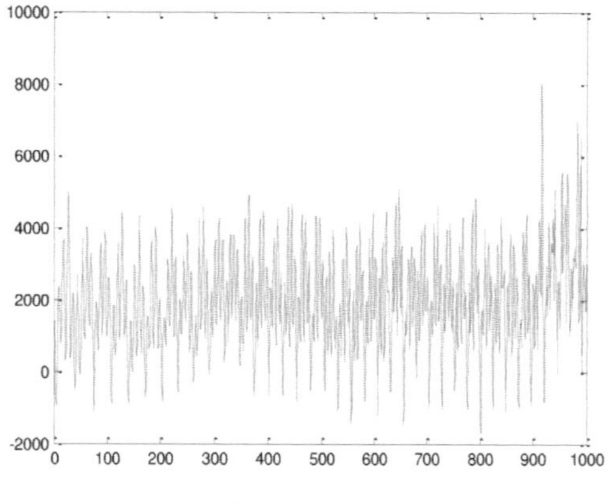

B : Fx of the WFT

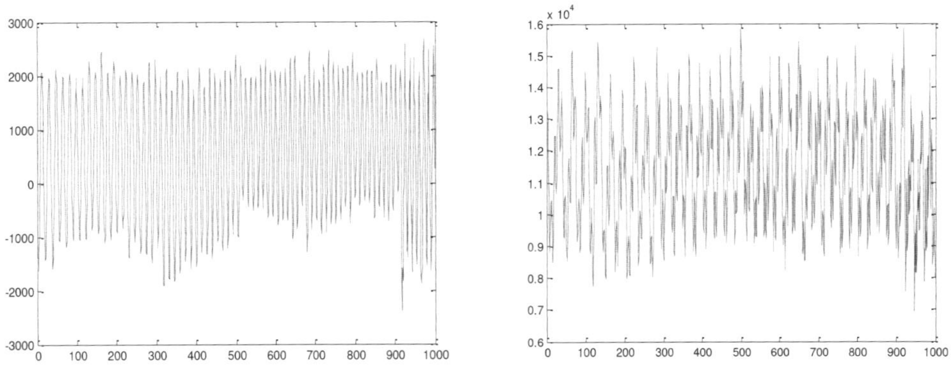

C : Fy of the WFT D : Fz of the WFT

Figure 2. The cobblestone road and the original load data

4. Experimental verification

It can be seen from the figure 2 that each group of load data contains 1000 acquisition points. The range of t can be obtained by combining the experiment and the stationary conditions, which is $1 \leq t \leq 1000$. The first item of the stationary conditions is $E|X_t|^2 < \infty$, therefore, if the inequality $\sum_1^{1000} E|X_t|^2 < \infty$ is established, $E|X_t|^2 < \infty$ naturally holds. $|X_1|^2$ of Fx is shown in figure 3, and it is easy to calculate $E|X_t|^2 = 1.205 * 10^{10}$, although it is large, but it is still a limited number. Therefore, the $\{X(t)\}$ of Fx satisfy the first item of the stationary conditions, Similarly available Fy and Fz also satisfy the first item.

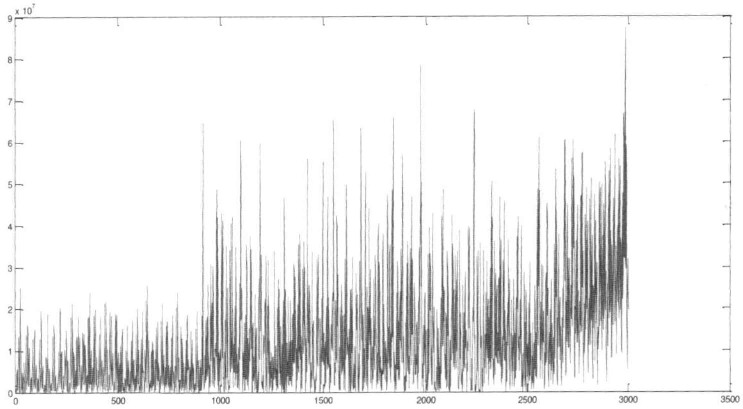

Figure 3.$|X_1|^2$ of Fx

The second item of the stationary conditions is $EX_t = m$, which means that the mean value of each acquisition point must be constant. Similarly, Fx was taken as an example to explain. EX_t = [EX_1 EX_2 … EX_{1000}] is shown in figure 4, and the fluctuation range of EX_t is equal to its maximum value of Ex minus its minimum value, which is 213.9N and is within the precision range of WFT System. It should be noted that the accuracy of the vertical force of the WFT system is 1% of its full scale [5], and the full scale of this WFT system is 40KN, so the error within 400N is acceptable and EX_t can be regarded as a constant.

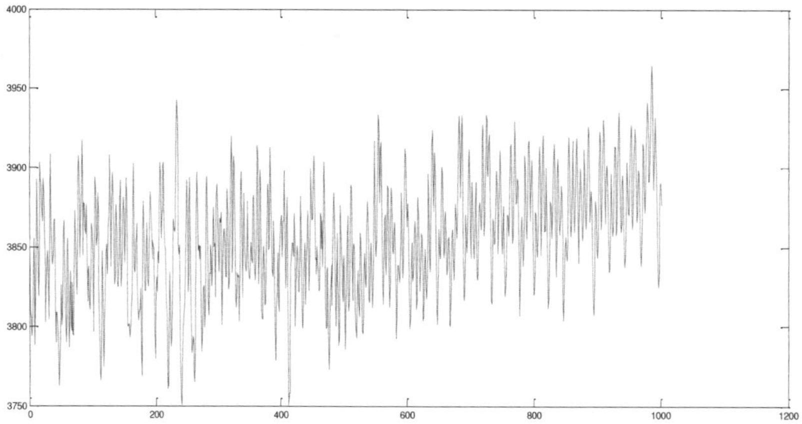

Figure 4. EX_t of Fx

The third condition of the stochastic process implies that the covariance function between any two moments of the stochastic process is independent of the time points and is only related to the distance between them. The sample autocorrelation function of Xt = (X_1 X_2 … X_{1000}) is defined as follow.

$$\gamma'(h) = n^{-1}\sum_{j=1}^{n-h}(x_{j+h} - x')(x_j - x'),\ 0 \le h \le n\ ,\ n = 1000.$$

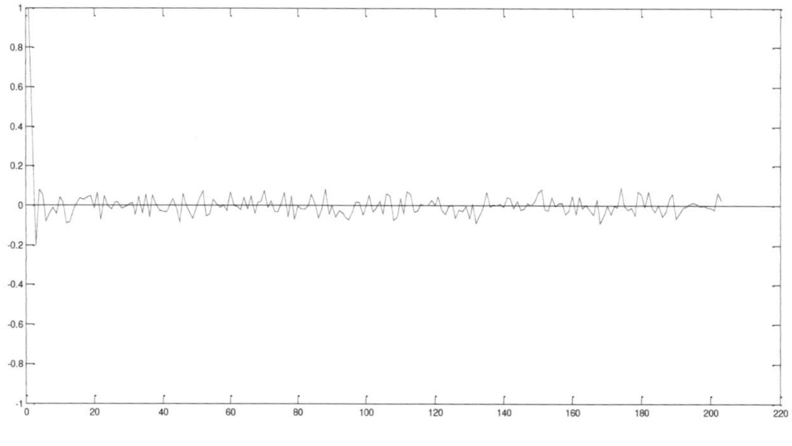

Figure 5. Sample correlation (coefficient) function ρ, $0 \le h \le 200$

According to the above formula is very convenient to calculate the sample autocorrelation coefficient (ρ) which is shown in figure 5 with h from 1 to 200. It is obviously to know that the value of ρ is less than 0.2 except when h equals 0 from the curve in figure 5, which means it is enough to get a conclusion that any two moments of Xt has no obviously relationship.

5. Summary

Although only making longitudinal force Fx as an example for a brief description, while it also applies to the lateral force Fy and the vertical force Fz. In this paper, the theoretical analysis and experimental verification show that the measured road load spectrum belongs to the stationary random sequence. So the road feature extraction and road type identification has a valid theoretical basis.

References

[1] Peter J.brockwell, Richard A.Davis. Time series: Theory and Methods. Springer. 2011,p8 –p23.
[2] Duan Huming, et al. Feature parameter extraction and statistical analysis of road surface measurement data. *Journal of Vibration and Shock*, 2013. 32(1): p30-34.
[3] Yong-chen, W.G, The methods of power spectrum estimation based on measured pavement. 2011, IEEE. p. 2787 - 2790.
[4] Hong Yongfu. Automobile development engineering. Machinery Industry Press.2014, p115-160.
[5] Lin, G., et al., A self-decoupled three-axis force sensor for measuring the wheel force. *Proceedings of the Institution of Mechanical Engineers, Part D: Journal of Automobile Engineering*, 2014. 228(3): p. 319-334.

IOP Conf. Series: Materials Science and Engineering **234** (2017) 012011 doi:10.1088/1757-899X/234/1/012011

Research on Aeroheating of Hypersonic Reentry Vehicle Base Flow Fields

Qin Xuguo[1], Shui Yongtao[1], Wang Yonghai[1], Chen Gang[1] and Li Qiang[1]

[1] Beijing Institute of Space Long March Vehicle; Beijing, China

by0405118@163.com

Abstract. The structure of the base flow of a hypersonic reentry vehicle and the resulting base pressure and heat transfer have been studied by numerical study. The compressible Navier-Stokes equations are solved by the finite-volume method. SST k-ω turbulence model is used, and comparisons are made with flight test. Attention was focused on assessing the effects of angle of attack and Mach number. It was found that angle of attack can significantly alter the wake flow structure and reentry vehicle base pressure and heating distributions. The results of the simulation may provide a theoretical basis for the design of the thermal protection system of hypersonic reentry vehicles.

1. Introduction

When the hypersonic flight vehicle reentr into the atmosphere, the problem of aerodynamic thermal protection has always been a problem to be considered in the hypersonic flight vehicle design. The high enthalpy flow field generated by hypersonic velocity will produce an extreme thermal environment. Thus, a thermal protection system must be designed.

The thermal environment at the bottom of the flight vehicle is lower than that of the tip and the large area. However, the material,size and weight of the bottom thermal protection system have a direct impact on the overall design, so the aerodynamic heating caused by the bottom flow has always been a hot topic. The prediction of the bottom thermal environment will directly affect the design of the hypersonic flight vehicle.

When the hypersonic flight vehicle reentry into the atmosphere, the bottom flow is very complex, including the mutual interference of shock wave, shear layer, recirculation zone, separation wake vortex and so on[1,2]. In the presence of an angle of attack, the structure of the wake flow field is tilted and the viscous shearing effect is enhanced, making the heat flow and pressure distribution different from the zero angle of attack. Most of the previous studies focused on the zero angle of attack[3-5], while few studies were about the angle of attack. When the experiment studies the bottom flow, its flow field is easy to be interfered and difficult to accurately measure owing to the impact of such support. Therefore, we can use wind tunnel experiment or flight test method to get the experimental data, conducting the CFD verification and then calculating the state that the experimental method cannot reach by adopting the CFD method.

In this paper, CFD method is used to simulate the distribution of bottom heat flow and pressure upon the reentry of hypersonic flight vehicle. The influence of the angle of attack and Ma number on the heat flow and pressure distribution is studied. The flow mechanism is also discussed. It provides supplementary reference for the wind tunnel experiment and the flight test of bottom heat flow.

2. Numerical Method

The simulation model is a simple cone. The length is L=3960mm. The radius of nose is 2.54mm, Ma_∞ =5,15,10,20,α=0°,5°,10°,15°,20°,25°,30°;P_∞=2801.55Pa,T_∞=221K. The altitude is H=24.38Km. The temperature of wall at base is T_{wb}=354K.

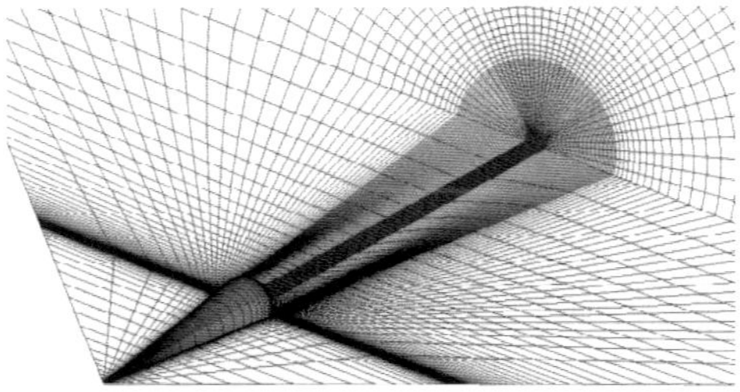

Figure 1. Three-dimensional grid for simulation.

Half model is used to be simulated, as shown in Figure 1. The vertex of nose is the origin of coordinates. The thickness of first layer of grid is 10^5 m. The overall three-dimensional multi-block grid consists of 1.6×10^5 points. The compressible Navier-Stokes equations are solved by the finite-volume method. SST k-ω turbulence model is used. Higher order Roe scheme is used, and the time term is solved by implicit scheme.

3. Computed result and analysis

3.1. Validation

The bottom flow of the hypersonic flight vehicle is complicated, which brings a lot of difficulties to the numerical simulation, so it is necessary to verify the numerical method. First, the flow field at Ma = 20, α = 0 ° is simulated and compared with the flight test results. As shown in Figure 2, the flight test values are more evenly distributed along the radial direction and the calculated heat flow peak value is bigger than the test value. The experimental results are in agreement with the results obtained in the literature [4] with DES method and literature [6] with RANS.

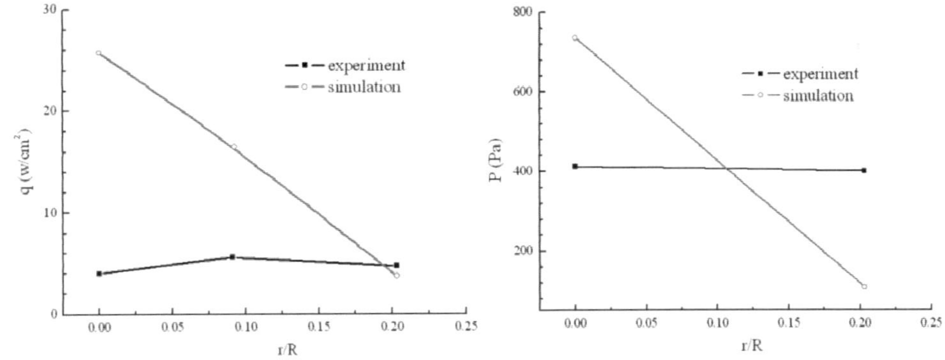

(a) Heat flux distribution (b) Pressure distribution

Figure 2. Comparison of calculated value and trial value.

3.2. Heat flux and pressure distribution with the effect of angle of attack

The effects of the angle of attack and Ma number on bottom heat flow and pressure are numerically simulated. The heat flow distribution on the bottom center line from the windward to the symmetry plane of the leeward side is shown in Figure 3. It can be seen that, at α= 0 ° , the heat flow peak value appears at the bottom of the center position. With the increase of the angle of attack, the heat flow peak value moves toward the leeward side. The greater the Ma number, the more obvious the peak value of bottom heat flow that changes with the angle of attack.

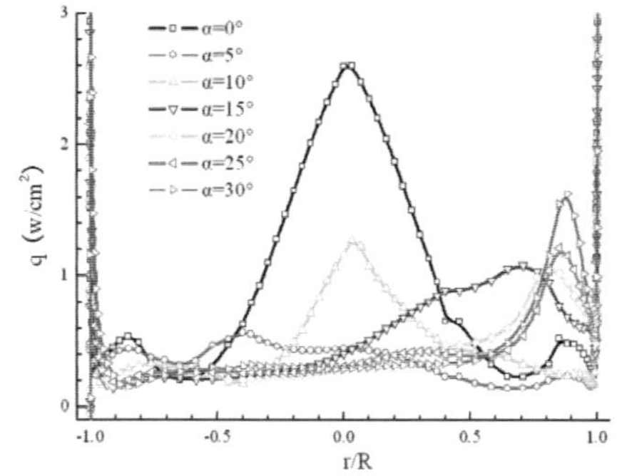

(a) Ma=5

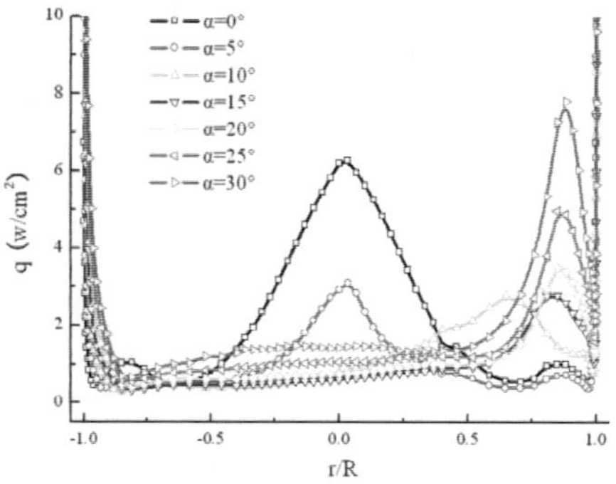

(b) Ma=10

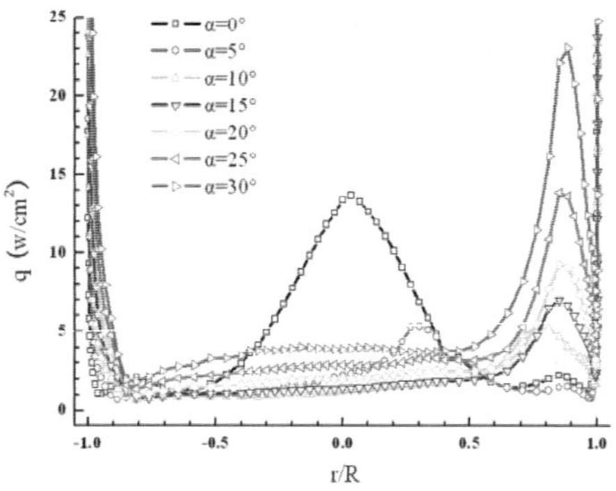

(c) Ma=15

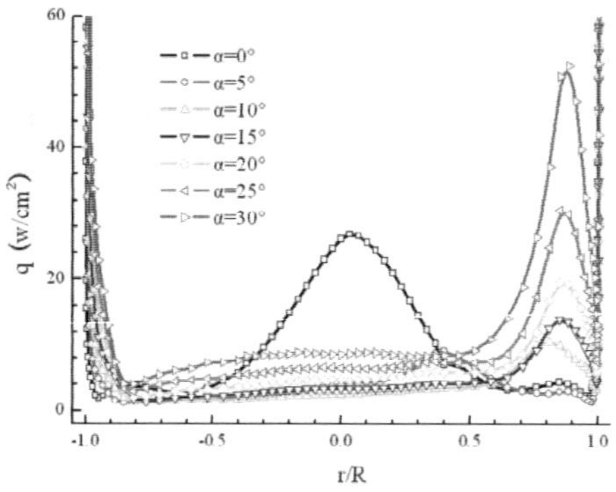

(d) Ma=20

Figure 3. Heat flux of center line with the effect of angle of attack.

With the increase of the angle of attack, the peak value of the bottom heat flow decreases first and then increases. The minimum peak value of the bottom heat flow appears at the angle of attack $\alpha = 5$ °. The peak value of bottom heat flow is smaller than that at $\alpha = 0$ ° when the angle of attack increases to $\alpha = 30$ ° for the smaller Ma number. With the increase of Ma number, the larger the angle of attack, the larger the increase of the peak value of the heat flow at such angle of attack. For example, when Ma = 15, the heat flow peak value at $\alpha = 25$ ° is equivalent to that at $\alpha = 0$ °, and when Ma = 20, the heat flow peak value at $\alpha = 25$ ° is higher than that at $\alpha = 0$ °. When Ma = 20, the heat flow peak value at $\alpha = 30$ ° is close to twice the peak value at $\alpha = 0$ °. Therefore, if the flight attitude is at large angle of attack when the vehicle flies at high Ma, the peak value of bottom heat flow at such angle of attack should be considered.

With the increase of Ma number, the bottom heat flow value at each angle of attack increases. From Ma = 5 to Ma = 20, the peak value of heat flow increases by 9.3 times at $\alpha = 0$ ° and 31.1 times at α=30°. For the higher Ma number, the location and magnitude of the peak value of the heat flow change with the angle of attack at a smaller angle of attack. As shown in Figure 3 (d), the peak value of heat flow from $\alpha = 0$ ° to $\alpha = 5$ ° decreases to about 1/3 at $\alpha = 0$ °, and the location varies from the center line r/R = 0 to r/R = 0.5. With the increase of the angle of attack, the location of the heat flow peak value varies little.

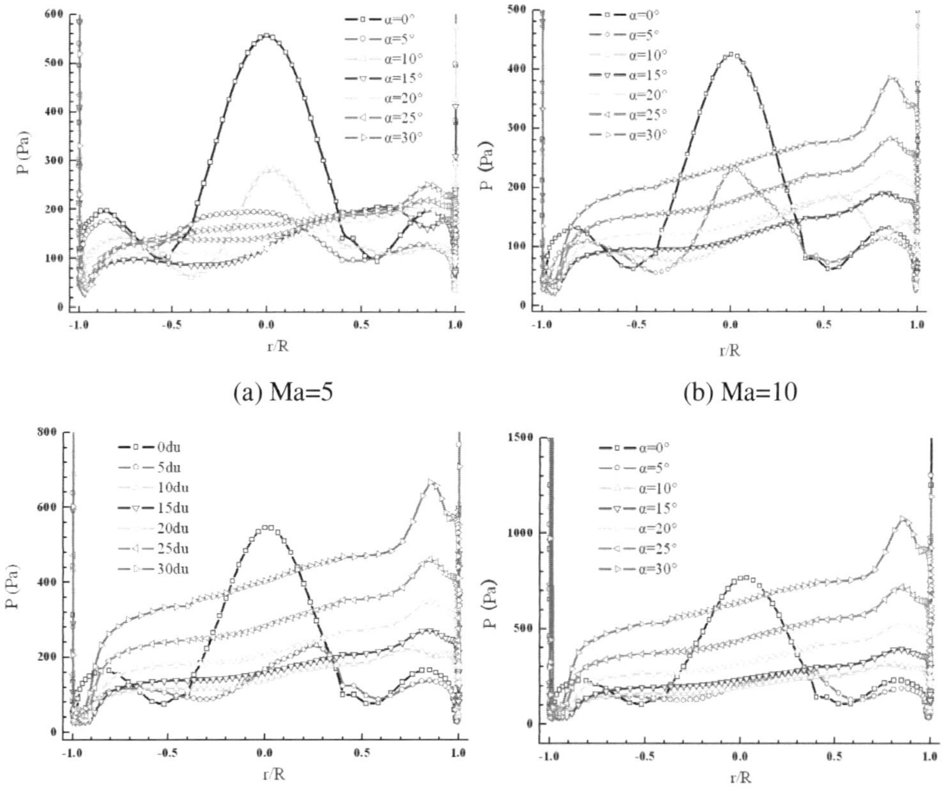

(a) Ma=5 (b) Ma=10

(c) Ma=15 (d) Ma=20

Figure 4. Pressure of center line with the effect of angle of attack.

The pressure distribution on the bottom center line is shown in Figure 4. The trend of the pressure and heat flow with the change of the angle of attack and Ma number is similar to the heat flow. When $\alpha = 0$ °, the peak value of pressure appears at the center of the bottom. When the angle of attack occurs, the pressure on the leeward side is reduced and the flow line on the conical surface flows along the cone to the leeward side. At the same time, the bottom stagnation point moves toward the leeward side. At high Ma, the pressure distribution along the centerline is more evenly distributed than the heat flow distribution.

3.3. The rule of the bottom flow field with the change of angle of attack and Ma number

The Ma distribution of the bottom flow field of the flight vehicle at Ma = 20 is shown in Figure 5, and the topology of the wake flow field at other Ma numbers is similar to that at Ma = 20. The wake flow field at the angle of attack $\alpha = 0$ ° is shown in Figure 5. It can be seen that there is a very large Prandtle-Meyer dilatational wave region in the shoulder region of the flow field at the bottom of the flight vehicle and a large recirculation zone at the bottom of the flight vehicle. The downstream flow field produces the recompression shock wave, and the reattachment point position is in the vicinity of the downstream $L / R \approx 1.49$. As the Ma number increases from 5 to 20, the position decreases to near $L / R \approx 1.26$. At the calculated angle of attack, no separation occurs on the leeward side of the flight vehicle. With the increase of the angle of attack, the shear effect of the bottom flow field is enhanced; the area of the bottom flow field is reduced; and the recirculation zone and the bottom stagnation point move toward the leeward side.

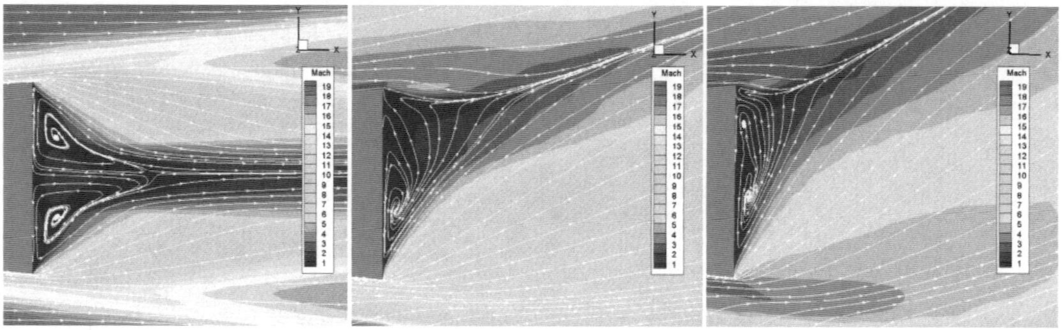

(a) α=0 ° (b) α=15 ° (c) α=30 °

Figure 5. Contour of Ma with the effect of angle of attack, Ma=20.

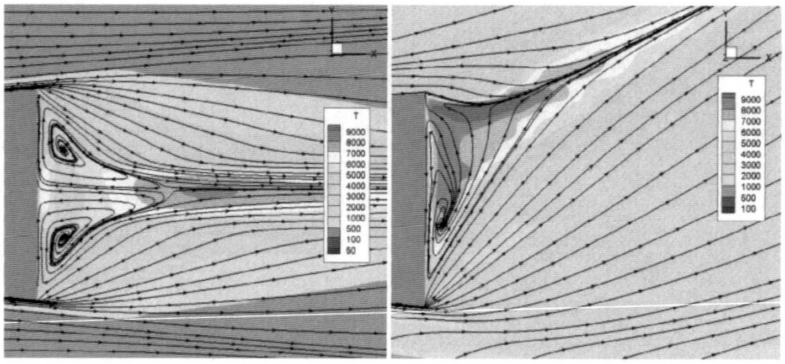

(a) α=0 ° (b) α=25 °

Figure 6. Contour of temperature with the effect of angle of attack, Ma=20.

The temperature distribution of the bottom flow field at Ma = 20 is shown in Figure 6. It can be seen that the temperature near the stationary point of the recirculation zone in the wake flow field is very high at α = 0°. With the increase of the angle of attack, the recirculation zone moves toward the bottom leeward direction and is closer to the wall, thus the bottom heat flow increases.

The distribution of the bottom heat flow with the change of the angle of attack at Ma = 20 is shown in Figure 7. It can be seen that the maximum value of the heat flow at α = 0 ° appears at the center of the bottom. As the angle of attack increases, the peak value of the heat flow moves toward the leeward side. However, at the same time, we can see that apart from the heat flow and pressure peak areas near the leeward side on the center line, there are two relatively concentrated areas of heat flow and pressure near the center line of the leeward side. When the angle of attack occurs, the pressure on the leeward side of the flight vehicle cone is reduced and the flow line on the conical surface flows along the cone to the leeward side, which changes the shape of the recirculation zone.

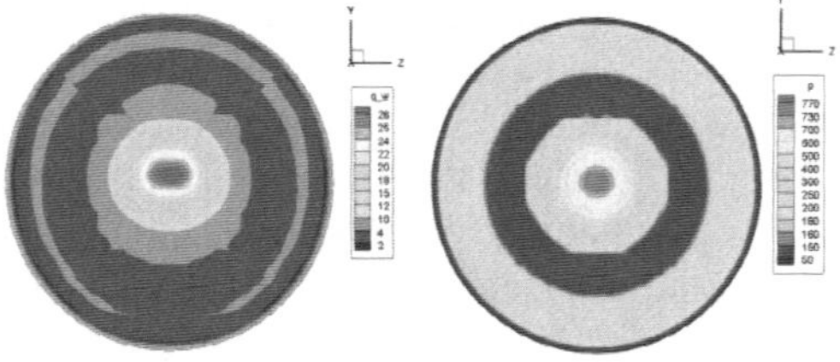

(1) α=0°

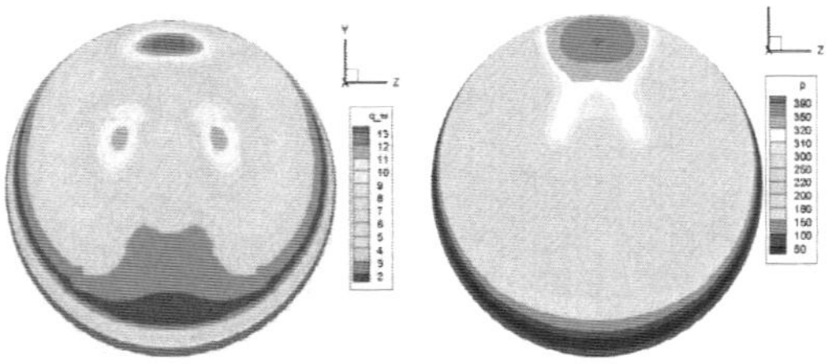

(2) α=15°

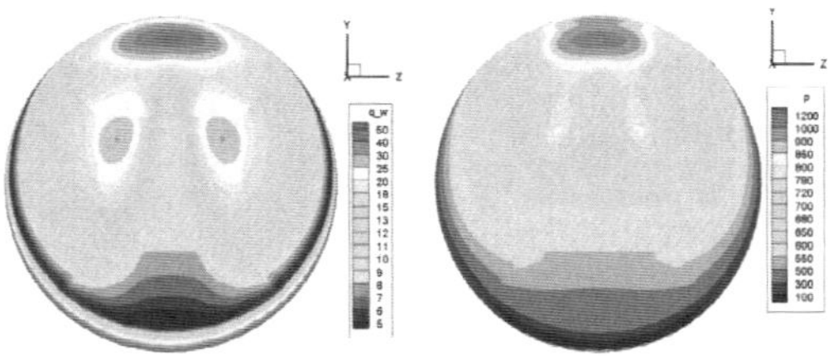

(3) α=30°

(a) Heat flux (b) Pressure

Figure 7. Contour of heat flux and pressure with the effect of angle of attack, Ma=20.

4. Conclusions

In this paper, the effects of the angle of attack and Ma number on the aerodynamic heating of the bottom flow field of a hypersonic flight vehicle are studied. The result shows:

(1) As the angle of attack increases, the peak value of the heat flow at the bottom of the flight vehicle moves toward the leeward side, and the peak value of heat flow decreases first and then increases.

(2) With the increase of Ma number, the bottom heat flow value at each angle of attack increases. The greater the Ma number, the more obvious the peak value of bottom heat flow that changes with the angle of attack.

(3) With the increase of the angle of attack, the shear effect of the bottom flow field is enhanced; the recirculation zone of the bottom flow field is reduced; and the recirculation zone and the bottom stagnation point move toward the leeward side.

Reference:

[1] J.Parker Lamb, William L. Oberkampf. Review and development of base pressure and base heating correlations in supersonic flow[J]. Journal of Spacecraft and Rockets, Vol.32, No.1, 1995:8-23.

[2] Datta V. Gaitonde. Mean flowfield structure of a supersonic three-dimensional base flow[J]. Journal of Aircraft, Vol. 47(2):368-382, 2010.

[3] Krishnendu Sinha, Effect of Reynolds Number on Detached Eddy Simulation of Hypersonic Base Flow[R], AIAA-2007-1457, 2007.

[4] Michael Barnhardt, Graham V. Candler. Detached eddy simulation of hypersonic base flows during atmospheric entry[R]. AIAA-2006-3575, 2006.

[5] Dillon, J. L, Carter, H. S. Analysis of Base Pressure and Base Heating on a 5° Half-Angle Cone in Free Flight Near Mach 20 (Reentry F), NASA-TM X-2468, 1972.

[6] T.C. Lin, L,K. Sproul. Hypersonic reentry vehicle wake flow fields at angle of attack[R]. AIAA-2006-582, 2006.

Study on BTT Coordinated Turn Autopilot Design for Reentry Gliding Vehicle

Li Qiang, Shui Yongtao, Wang Yonghai, Wang Gang and Qin Xuguo

No.1, Nandahongmen Road, Fengtai District, Beijing, China

Email:liqiang2568837@126.com

Abstract. Two-loop acceleration yaw-autopilot with a PI compensator and the engineering approximation of sideslip angle rate feedback for autopilot inner-loop are put forward accounted for BTT coordinated turn control of reentry gliding vehicle. The effect of turning acceleration to body-frame yaw channel is analyzed deeply, two-loop acceleration yaw-autopilot with a PI compensator is advanced to ensure system rapidity and stability based on minimum output by turning acceleration. The convergence essence of sideslip angle rate feedback is presented, furthermore, the engineering implementation composed by yaw angle rate and feedforward compensator is presented. Finally, the simulation results show that the autopilot design can make sideslip angle to be zero fast, improve the performance of coordinated turn. The design has certain robustness and application value.

1. Introduction

The importance of long-distance, fast and accurate attack is increased rapidly to fulfill requirement of modern war. So reentry gliding vehicle based on glide trajectory is widely concerned by countries[1]. Bank to turn (BTT) control technology must be applied because of high lift-to-drag configuration and maneuverable requirement of flight vehicle, and the strong three-channel coupled effect is produced by high rolling rate[2]. Pitch channel and roll channel are designed to respond command rapidly, and yaw channel must achieve coordinated turn control to eliminate sideslip angle and three-channel coupling, which realizes stable control of missile. So coordinated turn autopilot design of BTT missile is the key technology for control system.

Many experts have pay attention to the coordinated turn technology, and produce many important technical achievements. The autopilot design technology based on classical control theory has three class typical design methods. The first one is to eliminate sideslip by sideslip feedback. The second one is to eliminate sideslip by lateral acceleration feedback, because lateral acceleration of body-frame is produced by sideslip angle. The third one is keeping yaw angular rate unchanged to realize coordinated turn in constant roll angle and flight rate. The lateral acceleration feedback is most common and effective method because of simple system configuration and easy engineering application[3-5].

Based on the design philosophy of yaw acceleration feedback, in this paper the effect of autopilot configuration to coordinated turn performance is analyzed deeply, increasing open-loop gain is the key point to accomplish coordinated turn perfect, so two-loop acceleration yaw-autopilot with a PI compensator is put forward. With the comparison between different feedback information of inner-loop, it comes the conclusion that sideslip angle rate feedback is better than yaw angle rate. To meet engineering requirement, the engineering approximation of sideslip angle rate feedback is also presented.

2. Yaw channel aerodynamic analysis

During the tuning of flight vehicle, lateral turning acceleration of inertial coordinate system called a_{zd} is produced by the component of lift force in longitudinal-symmetrical plane. Speed vector is deflected by a_{zd}, and sideslip angle is produced. Based on the linear hypothesis, the dynamic equation of yaw channel is as follows[6].

$$\begin{cases} \ddot{\psi} = -a_\beta \beta - a_\omega \dot{\psi} - a_\delta \delta \\ \dot{\psi}_V = b_\beta \beta + b_\delta \delta - \dfrac{a_{zd}}{V} \\ \beta = \psi - \psi_V \\ a_{zV} = -V \dot{\psi}_V \end{cases} \quad (1)$$

Where ψ, $\dot{\psi}$ and $\ddot{\psi}$ are yaw angle, rate and acceleration, ψ_V and $\dot{\psi}_V$ are heading angle and rate, δ is yaw rudder deflection angle, V is flight speed, β is sideslip angle, $a_\beta, a_\omega, a_\delta, b_\beta, b_\delta$ are the aerodynamic derivatives of yaw channel defined in reference[2]

The overall system dynamics is given by,

$$\begin{bmatrix} \dot{\psi} \\ \ddot{\psi} \\ \dot{\psi}_V \end{bmatrix} = \begin{bmatrix} 0 & 1 & 0 \\ -a_\beta & -a_\omega & a_\beta \\ b_\beta & 0 & -b_\beta \end{bmatrix} \begin{bmatrix} \psi \\ \dot{\psi} \\ \psi_V \end{bmatrix} + \begin{bmatrix} 0 & 0 \\ -a_\delta & 0 \\ b_\delta & -\dfrac{1}{V} \end{bmatrix} \begin{bmatrix} \delta \\ a_{zd} \end{bmatrix} \quad (2)$$

Considering the requirement of information measure and coordinated turn research, and defining system output are a_{zb}, $\dot{\psi}$ and β, a_{zb} is yaw acceleration, the system dynamics is

$$\begin{bmatrix} a_{zb} \\ \dot{\psi} \\ \beta \end{bmatrix} = \begin{bmatrix} -b_\beta V & 0 & b_\beta V \\ 0 & 1 & 0 \\ 1 & 0 & -1 \end{bmatrix} \begin{bmatrix} \psi \\ \dot{\psi} \\ \psi_V \end{bmatrix} + \begin{bmatrix} -b_\delta V & 0 \\ 0 & 0 \\ 0 & 0 \end{bmatrix} \begin{bmatrix} \delta \\ a_{zd} \end{bmatrix} \quad (3)$$

Apparently a_{zb} is not correlative with a_{zd}, but ψ_V is varied because of a_{zd}, a_{zd} can influence a_{zb} also. The transfer function a_{zb}/a_{zd} and $\dot{\psi}/a_{zd}$ are as follows.

$$\frac{a_{zb}}{a_{zd}} = \frac{-b_\beta(s+a_\omega)}{s^2+(b_\beta+a_\omega)s+(a_\omega b_\beta+a_\beta)} = \frac{-b_\beta(s+a_\omega)/\omega_m^2}{s^2/\omega_m^2+2\mu_m s/\omega_m+1} \quad (4)$$

$$\frac{\dot{\psi}}{a_{zd}} = \frac{-a_\beta/V}{s^2+(b_\beta+a_\omega)s+(a_\omega b_\beta+a_\beta)} = \frac{-a_\beta/(V\omega_m^2)}{s^2/\omega_m^2+2\mu_m s/\omega_m+1} \quad (5)$$

Where $\omega_m = \sqrt{a_\beta + a_\omega b_\beta}$ is undamped natural frequency, $\mu_m = (b_\beta + a_\omega)/(2\sqrt{a_\omega b_\beta + a_\beta})$ is damping radio. Steady-state a_{zb} and $\dot{\psi}$ are as follows.

$$a_{zb} = -b_\beta a_\omega a_{zd}/(a_\omega b_\beta + a_\beta) = -b_\beta a_\omega a_{zd}/\omega_m^2 \quad (6)$$

$$\dot{\psi} = (-a_\beta/V)a_{zd}/(a_\omega b_\beta + a_\beta) = (-a_\beta/V)a_{zd}/\omega_m^2 \quad (7)$$

If static stability is lower, ω_m is smaller, a_{zb} and $\dot{\psi}$ during turning are bigger, β is bigger, which is not benefit to coordinated turn.

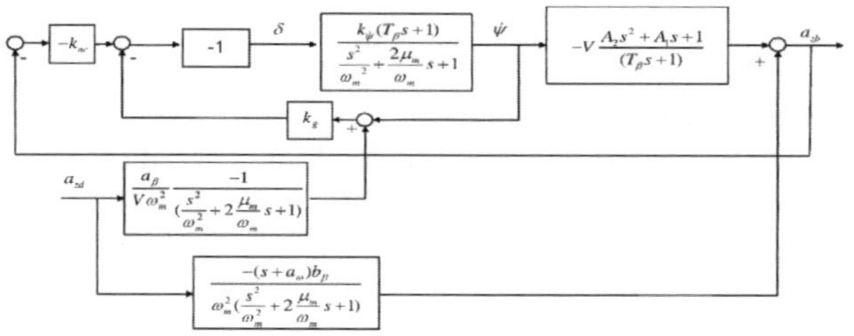

Figure1. A typical block diagram for coordinated turn autopilot design

The command is zero for coordinated yaw autopilot design all the time, a_{zb} and $\dot{\psi}$ produced by a_{zd} are disturbance to yaw autopilot. So the mission of autopilot design can be described that minimizing the output of a_{zb} when there is disturbance input of a_{zd}. The block diagram is shown in Fig.1, where k_{ac} and k_g are design parameters of autopilot, $A_1, A_2, T_\beta, k_{\dot{\psi}}$ are the aerodynamic derivatives of yaw channel defined in reference [2].

3. Coordinated turn autopilot structure research

Switching the yaw-autopilot structure based on Fig.1, a block diagram for two-loop acceleration autopilot with input a_{zd} is shown in Fig.2.

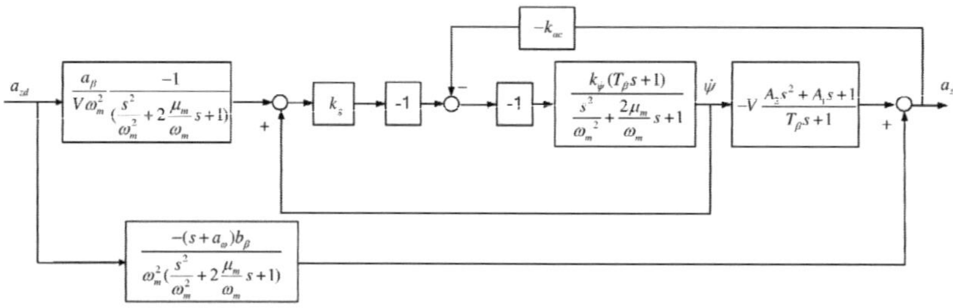

Figure2. The block diagram for two-loop acceleration autopilot with input a_{zd}

The transfer function a_{zb}/a_{zd} is as follows

$$a_{zb}/a_{zd} = k_g k_{\dot{\psi}}\left(\frac{s}{a_\beta^2 k_g k_{\dot{\psi}}}+1\right) \Big/ \left(\frac{s^2}{a_\beta}+\frac{b_\beta}{a_\beta}s+1\right)(1-k_{ac}k_{\dot{\psi}}V-k_g k_{\dot{\psi}}) \tag{8}$$

Defining $K_{a_{zb}/a_{zd}}$ is steady state yaw acceleration of per lateral turning acceleration, so smaller $K_{a_{zb}/a_{zd}}$ represents better coordinated turn performance. $K_{a_{zb}/a_{zd}}$ is given by

$$K_{a_{zb}/a_{zd}} = k_g k_{\dot{\psi}} / (1-k_{ac}k_{\dot{\psi}}V-k_g k_{\dot{\psi}}) \tag{9}$$

The open-loop gain of autopilot is given by

$$K_G = k_{ac}k_{\dot{\psi}}V / (k_g k_{\dot{\psi}}-1) \tag{10}$$

Substitution of expression for K_G into equation (10) yields

$$K_{a_{zb}/a_{zd}} = k_g k_{\dot{\psi}} / (K_G+1)(1-k_g k_{\dot{\psi}}) \tag{11}$$

Where $k_{\dot{\psi}}$ is nature ability of flight vehicle, and adjusting k_g to reduce $K_{a_{zb}/a_{zd}}$ is also limited by $k_{\dot{\psi}}$ evidently. So the most direct and effective way is increasing K_G, which means that increasing open-loop gain is most benefit to improve coordinated turn ability.

With open-loop gain of 0.49 and 0.98 respectively, the time domain and frequency domain results are given in Fig.3 and Fig.4 for 1g input a_{zd}

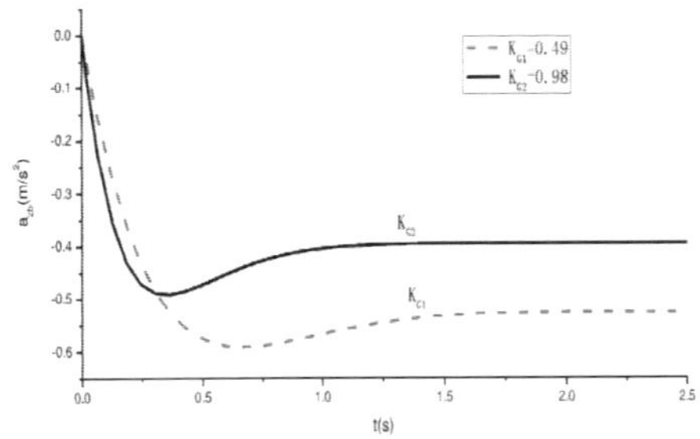

Figure 3. Time domain curves with K_G

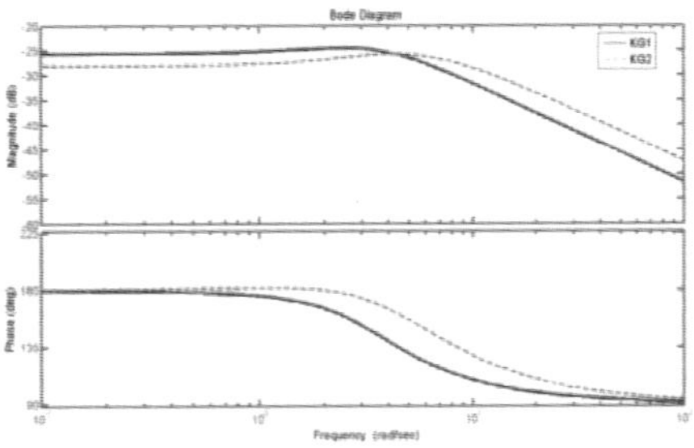

Figure 4. Frequency domain Bode curves with K_G

Fig.3 displays the steady-state disturbance output can be reduced by increasing K_G. Fig.4 shows that system low-frequency gain degenerated obviously, which promotes the coordinated turn ability, but can not make disturbance output to be zero. The disturbance output is eliminated drastically required too large open-loop gain, which will make the system unstable. So the essential method is to add a proper PI compensator in forward channel, which can make low-frequency gain degenerated and maintain system stable[7-8].

With a PI compensator in forward channel, the block diagram is shown in Fig.5.

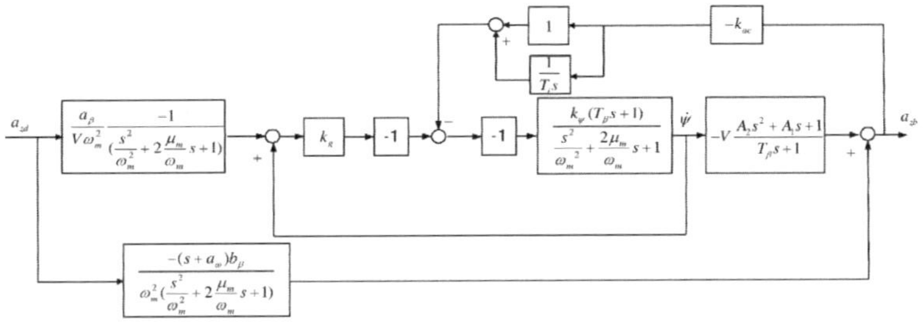

Figure5. The block diagram for two-loop acceleration autopilot with a PI compensator

Where T_i is compensator time constant. With open-loop gain of 0.49 for two-loop acceleration autopilot with a PI compensator, the time domain and frequency domain results are given in Fig.6 and Fig.7 for 1g input a_{zd}.

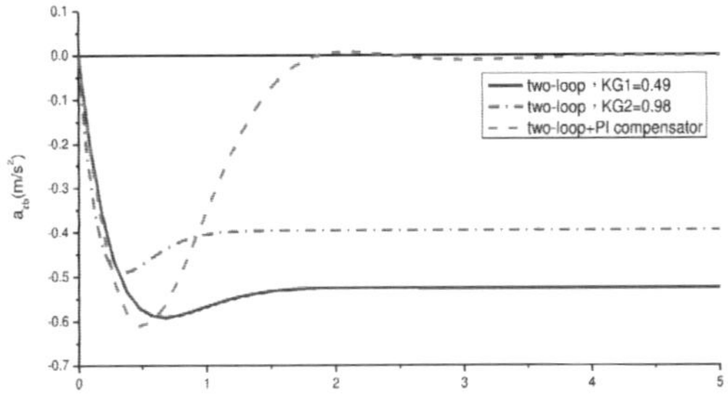

Figure6. Responses with or without PI compensator

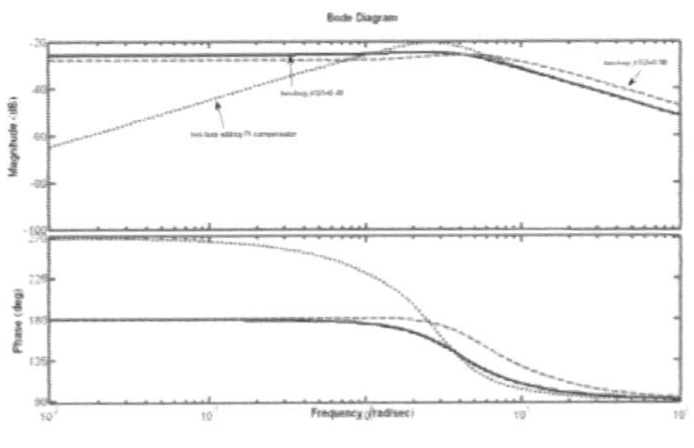

Figure7. Bode curves with different open-loop gains

Fig.6 and Fig.7 display the proper PI compensator can make low-frequency gain degenerated, which will make yaw-acceleration caused by lateral turning acceleration to be zero quickly and maintain system stable. So the two-loop acceleration autopilot with a PI compensator is much better to accomplish coordinated turn.

4. Inner-loop feedback characteristic

Generally, inner-loop of yaw-autopilot is designed to improve system damp and yaw angle rate is selected as feedback information. However, there is less research on the effect of inner-loop feedback to coordinated turn with input lateral turning acceleration. In steady-state of Eq.(7) $\dot{\psi}=(-a_\beta/V)a_{zd}/(a_\omega b_\beta+a_\beta)\approx -a_{zd}/V=\dot{\psi}_V$, the disturbance input of yaw angle rate exist all the time during sustained turning. The sideslip angle can be zero because of $\dot{\psi}\approx\dot{\psi}_V$ approximately.Sideslip angle is present from Eq. (1) as follow.

$$\dot{\beta}=[-b_\beta \quad 1 \quad b_\beta][\psi \quad \dot{\psi} \quad \psi_V]^T+[-b_\delta \quad \frac{1}{V}][\delta \quad a_{zd}]^T \tag{12}$$

Transfer functions $\dot{\beta}/a_{zd}$ and $\dot{\beta}/\delta$ can be written as follows.

$$\dot{\beta}/a_{zd}=s(s+a_\omega)/V\omega_m^2(s^2/\omega_m^2+2\mu_m s/\omega_m+1) \tag{13}$$

$$\dot{\beta}/\delta=(k_\beta B_1 s^2+k_\beta s)/(s^2/\omega_m^2+2\mu_m s/\omega_m+1) \tag{14}$$

Where k_β, B_1 are the aerodynamic derivatives of yaw channel defined in reference[2]. The conclusion that sideslip angle comes to zero with input turning acceleration in steady-state is clarified again. Comparing with inner-loop yaw angle rate feedback, we can draw the conclusion from Fig.8 that taking sideslip angle rate as feedback information is more propitious to eliminate yaw acceleration.

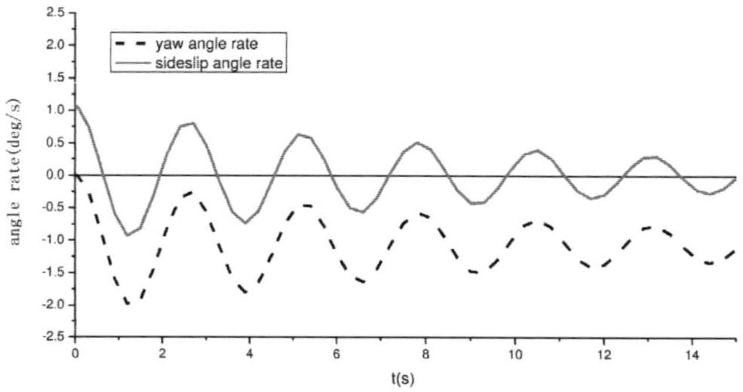

Figure8. Compare of inner-loop yaw and sideslip angle rate

5. Engineering approximation for sideslip angle rate inner-loop

Inner-loop sideslip angle rate feedback is more effective to coordinated turn, but the problem is how to measure sideslip angle rate accurately, the common method sideslip angle differential coefficient can bring serious noise problem. Taking channel-coupling into consideration, in the yaw channel there is [9-10]

$$\dot{V}_{zb}+V_{yb}\omega_{xb}-V_{xb}\omega_{yb}=F_{cz}/m \tag{15}$$

Where V_{xb}, V_{yb}, V_{zb} are the velocity in body coordinate system, ω_{xb}, ω_{yb} are roll and yaw angle rate, F_{cz} is total force of yaw channel, m is mass of flight vehicle.

$$\dot{V}_{zb}/V_{xb}=-V_{yb}\omega_{xb}/V_{xb}+\omega_{yb}+F_{cz}/mV_{xb} \tag{16}$$

There are the equations that $\beta=V_{zb}/V_{xb}$ and $\tan\alpha=-V_{yb}/V_{xb}$ approximately, where $F_{cz}=F_{czb}+mg\cos\vartheta\sin\gamma$ contains aerodynamic force and gravity. Because the requirement of coordinated turn, the aerodynamic force $F_{czb}=0$. g represents gravitational acceleration, ϑ, γ are pitch and yaw attitude angle, and make a hypothesis on $V_{xb}=V$, so form Eq. (16) we have the inner-loop feedback as

$$\dot{\beta} = \omega_{yb} + \omega_{xb} \tan\alpha + g\cos\vartheta \sin\gamma / V \tag{17}$$

Sideslip angle rate $\dot{\beta}$ is composed of three parts which can be obtained from inertial navigation system all , ω_{yb} is yaw angle rate and can be measured directly, $\omega_{xb}\tan\alpha$ is kinematic compensator and can be obtained by measuring roll angle rate and calculating attack angle, $g\cos\vartheta\sin\gamma/V$ is gravitational compensator and can be obtained by measuring attitude angle and velocity. So with kinematic and gravitational compensator, sideslip angle rate feedback is produced for the engineering implementation, which makes the performance of yaw-autopilot coordinated turn to be best.

With the consider of channel coupling, the block diagram of acceleration yaw-autopilot is described in Fig.9, the autopilot adopts two-loop system structure with a PI compensator, inner-loop sideslip angle rate feedback is produced from yaw angle rate adding kinematic and gravitational compensator.

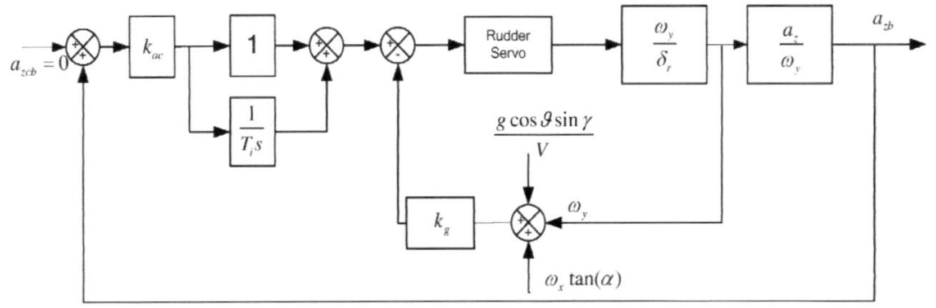

Figure9. The block diagram of acceleration yaw-autopilot

The yaw channel command is $a_{zbc}=0$, the pitch channel command is $\alpha_c = 10^{\circ}$, and the roll channel command is $\gamma_c = 45^{\circ}$. The yaw acceleration and rudder deflection angle with different inner-loop feedback are shown in Fig.10 and Fig.11.

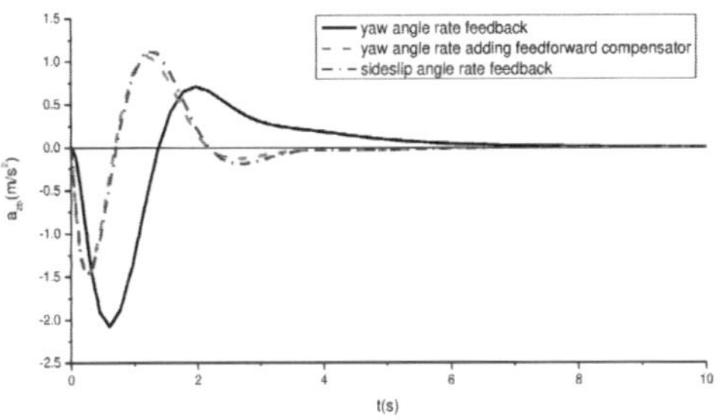

Figure10. The yaw acceleration

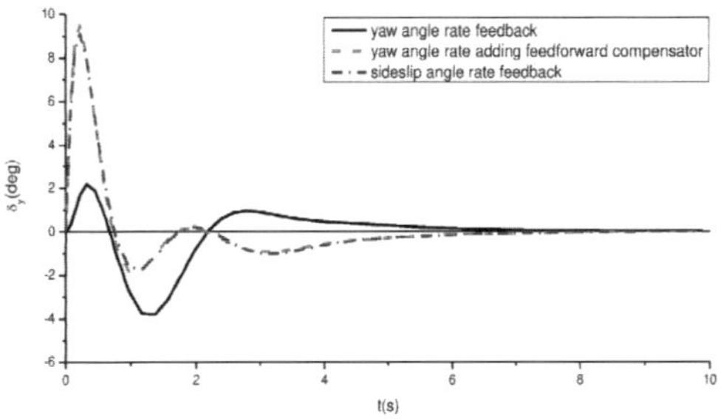

Figure11. The yaw rudder deflection angle

Based on Fig.10 and Fig.11, the feedback of yaw angle rate adding kinematic and gravitational compensator can decrease the maximum of yaw acceleration and make yaw acceleration to zero fast, which is much better for coordinated turn to eliminate sideslip angle, and the effect is equal to sideslip angle rate feedback. Comparing with sideslip angle rate feedback, the feedback of yaw angle rate adding kinematic and gravitational compensator need more rudder resource, which is used to eliminate sideslip angle quickly, so the sufficient rudder resource is the key point to coordinated turn.

6 .Conclusions

Based on the detailed performance analysis on yaw acceleration output by autopilot structure, two-loop acceleration yaw-autopilot with a PI compensator has been presented. The convergence essence of yaw acceleration by inner-loop sideslip angle rate feedback has been validated, and the engineering approximation of sideslip angle rate feedback has been confirmed, the coordinated turn yaw-autopilot structure has been presented. The simulation results show that the yaw-autopilot has been presented in this paper can eliminate the sideslip and channel coupling effectively, improve the coordinated turn ability clearly. The research in this paper has certain application value for control system design of reentry gliding vehicle.

References

[1] Li Yu. Study of Trajectory Optimization and Guidance Algorithm for Boost-glide Missile[D]. Harbin: Harbin Institute of Technology,2009.
[2] He Jing. Key Technology Research on Guidance and Control of Bank-to-turn Missile with Non-circular Cross Section in Terminal Guidance Phase[D]. Beijing: Beijing Institute of Technology,2012.
[3] Sun Baocai. Research on BTT Autopilot Design for Cruise Missile[D]. Beijing: Beijing Institute of Technology,2007.
[4] Arrow A. An Analysis of Aerodynamic Requirem ents for Coordinated Bank-to-Turn Autopilots[R]. NASA-CR-3644, Nov. 1982.
[5] Williams D E，Friedland B. Modern Control Therory for Design of Autopilots for Bank-to-Turn Missiles[J]. J.Guidance. 1987，10(4)：378~386.
[6] P Garnell, Guided Weapon Control Systems[M].London: Royal Military College of Science Press, 2004:161-172.
[7] Paul Zarchan. Tactical and Strategic Missile Guidance[M]. Virginia: American Institute of Aeronautics and Astronautics Inc, 2007:498-507.
[8] Ming Xin, S N Balakrishnan, Donald T S, et al. Nonlinear Bank-to-Turn/Skid-to-Turn Missile Outer-Loop/Inner-Loop Autopilot Design With $\theta-D$ Technique[C]. AIAA Guidance, Navigation, and Control Conference and Exhibit. Austin, Texas, August 11-14 2003.
[9] John R. Cleminson, Rick A. Hyde. Control and Maneuverability of a Square Cross-Section Missile[J]. Journal of Guidance, Control and Dynamics, 2006, 19(1):25-38.
[10] Tong Chunxia, Wang ZhengJie, Zhang Tianqiao. Decoupling System Design Based on Variable Structure System for BTT Missile[J]. Journal of Astronautics, 2006, 27(1):27-30.

IOP Conf. Series: Materials Science and Engineering **234** (2017) 012013 doi:10.1088/1757-899X/234/1/012013

Analysis of the Thermal Characteristics of Machine Tool Feed System Based on Finite Element Method

Xiaobo Mao[1], Kuanmin Mao[1], Yikang Du[1], Fengyun Wang[1] and Bo Yan[1]

[1]School of Mechanical Science and Engineering Huazhong University of science and Technology, Wuhan, 430074, China

kmmao4645@sina.com, maoxiaobo68@hust.edu.cn

Abstract. The loading of mobile heat source and boundary conditions setting are difficult problems in the analysis of thermal characteristics of machine tools. Taking the machine tool feed system as an example, a novel method for loading of mobile heat source was proposed by establishing a function which was constructed by the heat source and time. The convective heat transfer coefficient is the key parameter of boundary conditions, and it varies with the temperature. In this paper, a model of "variable convection heat transfer coefficient" was proposed, and the setting of boundary conditions of thermal analysis was closer to the real situation. Finally, comparing results of above method and experimental data, the accuracy and validity of this method was proved, meanwhile, the simulation calculation and simulation time was reducing greatly.

1. Introduction

In the total error of modern machine tool manufacturing, the thermal error caused by thermal deformation is as high as 40%~70% [1]. Thermal analysis is an effective method to reduce the thermal error, but the accuracy of this method depends on a high precision temperature field model [2, 3]. In addition, the boundary conditions are complicated problems for thermal analysis of machine tools, so the setting of boundary conditions are often inconsistent with the actual situation, especially the key parameter of convection heat transfer coefficients (CHTCs). The CHTCs varies with temperature. It is obviously not correct to use a fixed CHTCs value in the thermal analysis. However, it is very difficult to calculate CHTCs exactly [4]. Yang [5] designed a new algorithm of mobile heat source loading by APDL language, and programmed the calculation procedures. Using finite element method, a process of twin-wire submerged arc welding was performed and the rule of calculation and loading of mobile heat source was studied by Tang [6]. But both of the above methods had the large amount of calculation and complex programming. Zhao.[7] proposed a method for computing the CHTCs of spindle surface, which could make the CHTCs of spindle surface as the CHTCs of a flat plate when air flows along the spindle surface. A calculation method of CHCTs of spindle system based on RBF neural network was presented [8], and the accuracy of calculation was improved. However, they only gave the calculation method of the CHCTs, did not propose the setting method of CHCTs in the thermal analysis under different temperature.

As mentioned above, these methods of calculation and the setting of boundary conditions are not ideal. In this paper, a new method was proposed to simulate the mobile heat source, which could greatly reduce the amount of computation and simulation time. As for the setting of CHCTs in the thermal analysis, the authors presented a method which could set the variable CHCTs under different temperature in the thermal analysis. Furthermore, the setting of boundary conditions was closer to the actual situation, and it could improve the accuracy of the simulation analysis. An experiment was carried out in the feed system of a milling machine.

2. Thermal test of a feed system

The feed system of the milling machine consists of a table, a saddle, and a ball screw. To accurately establish the temperature field of the feed system, nine platinum resistance temperature sensors (T1-T9) were fixed in the feed system, as shown in figure 1. The ambient temperature of the machine tool is 11.2 °C.

For thermal analysis of the feed system, the authors designed the following experiment: The feed speed of the table along the X axis was 4000 millimeter per minute, and the motor speed was 267 turns per minute. The total experiment time was four hours.

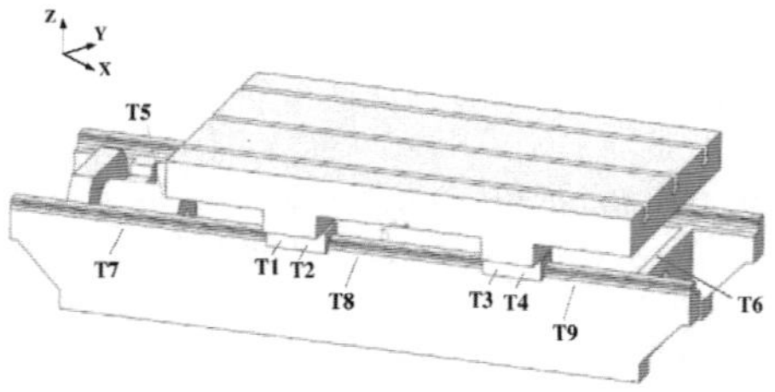

Figure 1. Solid model of the feed system.

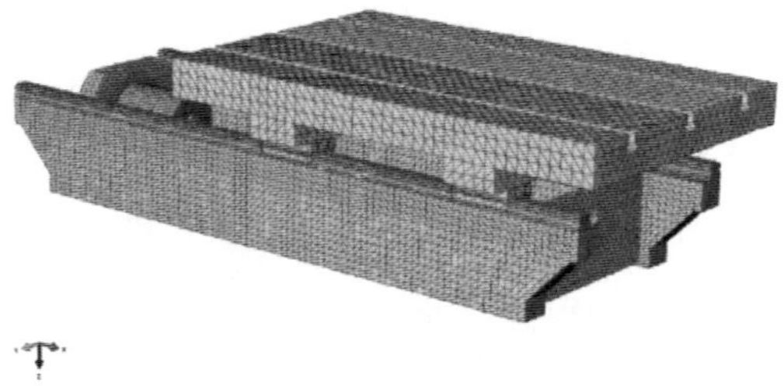

Figure 2. The FE model of the feed system.

3. Development of the FE simulation model

The temperature of the ball screw nut cannot be measured directly for the research of the feed system. However, two temperature sensors (T5 and T6) were fixed on near both ends of the ball screw, so the ball screw was not established in the solid model of the feed system. The finite element model was obtained with software of the ABAQUS, as shown in figure 2.

3.1. Determination of the material properties

There are two major materials in the feed system. Uniloy alloy was applied to guide ways, and carbon steel was applied to saddle and table. The two material properties were assigned them respective structures in the ABAQUS, and values of the properties are shown in table 1.

Table 1. Properties of the material.

Material	Uniloy Alloy	Carbon Steel
Density (Kgm^{-3})	7820	7840
Modulus of elasticity (Gpa)	206	200
Poisson's ratio	0.3	0.3
Linear expansion coefficient ($10^{-5}K^{-1}$)	1.1	0.85
Thermal conductivity ($Wm^{-1}K^{-1}$)	16.3	49.8
Specific heat capacity ($Jm^{-1}K^{-1}$)	460	465

3.2. Thermal sources analysis

The thermal sources of the feed system was from the motor, guide ways and the bearing. The heat generation of the guide ways, motor and bearing could be calculated by the following equation [9, 10].

$$Q = \mu F v \tag{1}$$

$$H = \frac{M_T n_1}{9550}(1-\eta) \tag{2}$$

$$H_f = 1.047 \times 10^{-4} n_2 M \tag{3}$$

Where M_T is the output torque of the motor, n_1 is the rotating speed of the motor, H_f is the heat generated power (w), n_2 is the rotating speed of the bearing (*rpm*), M is the frictional torque of the bearing (*Nmm*).

4. Heat source loading and bounding condition setting

4.1. Mobile heat source block loading

Thermal load could be calculated by equation (1)-(3), and it was loaded into the structural analysis in the form of heat flux. In this paper, the friction between the guide ways and the slider was mobile heat source. The amount and time of the simulation were greatly increasing if the thermal source was loaded by the current method of mobile heat source. Consequently, a new method for the loading of mobile heat source was proposed by establishing a function which was constructed by the heat source and time. The "table" commend of "Amplitude" in the ABAQUS was used to establish the function. To achieve the purpose of mobile heat source loading, the segmentation of heat flux and thermal loading schedule were designed. The heat flux density and the section loading time were shown in figure 3 and table 2.

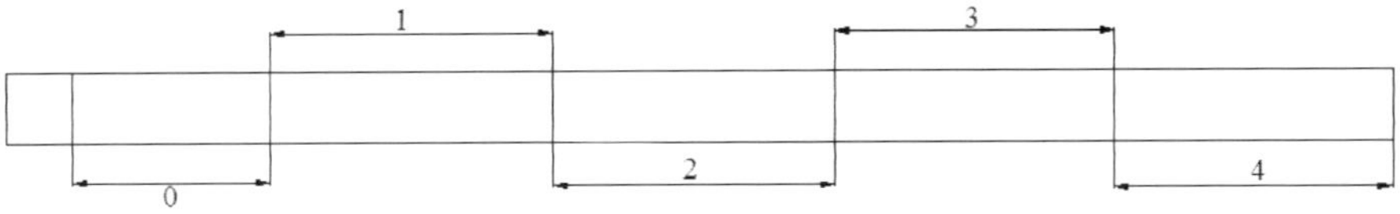

Figure 3. The segmentation of heat flux.

Table 2. The segmentation of thermal loading schedule.

Time(s) / Parts	1.5	3	4.5	6	7.5	9	10.5	12
1	■							■
2		■					■	
3			■			■		
4				■	■			

4.2. Variable heat transfer coefficient setting

The CHCTs was varying with temperature, but in the process of the thermal analysis, the temperature was changing with time. It was obvious that the analysis error had increased if the fixed CHCTs was used in thermal analysis. Therefore, the temperature was divided several intervals in the thermal analysis, and different CHCTs set up in each section of the temperature range in this paper. Then, a modal of variable CHCTs was established, and the method was closer to the reality and could improve the precision of thermal analysis. As for the variable CHTCs, the setting of them could be realised by ABAQUS, which was shown in figure 9.

Taking the cylindrical surface of the motor end connector as an example, the temperature interval was set to 2 °C. The CHTCs was calculated respectively when the temperature T_m was taking different values, such as 12 °C, 13 °C, 14 °C, 15 °C and 16 °C.

MATLAB was used to calculate the values of five CHCTs (Δ = 2 °C, 4 °C, 6 °C, 8 °C, and 10 °C), which were corresponding five temperature intervals in the temperature range of 13 °C ~ 21°C. The

IOP Conf. Series: Materials Science and Engineering **234** (2017) 012013 doi:10.1088/1757-899X/234/1/012013

values of the five CHTCs could be seen in table 3. Hence, with the help of the "interaction" module in the Abaqus, the setting of variable CHCTs was realizing, as shown in figure 9.

Table 3. Convective heat transfer coefficients (CHTCs).

Wall Temperature (T)	CHTCs (h)
$13°C$	3.42
$15°C$	4.08
$17°C$	4.5
$19°C$	4.8
$21°C$	5.1

5. Thermal simulation of the feed system

Sequentially coupled thermal stress analysis was performed by ABAQUS. The temperature distribution of the feed system was obtained by using "heat transfer" module, which was shown in figure 4. The simulated values of temperature measuring of the feed system were shown in figure 6. After that, the temperature field was loaded into the solid model as a predefined load, meanwhile, the "variable boundary conditions" was set using the "static general" module. The thermal deformation of the feed system was shown in figure 5.

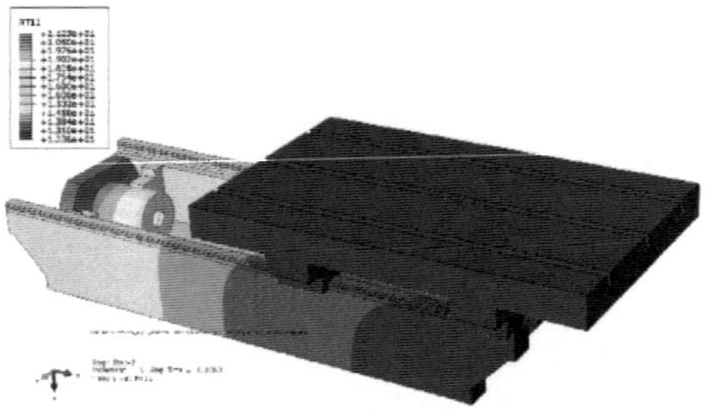

Figure 4. Temperature distribution of the feed system.

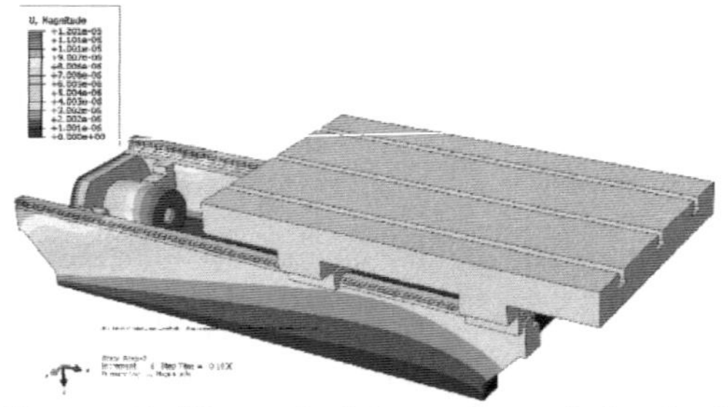

Figure 5. Thermal deformation of the feed system.

It could be seen that motor heat generation occupies the main position in heat source of the machine tool feed system from the figure 4 and figure 5. And the same time, the sliding friction heat between saddle and slider was the second heat source in the total heat source of the feed system. As a result, thermal deformation of the machine feed system was bigger than that the bearing end of the motor. This phenomenon was not only related to the characteristic of the machine tool itself, but also related to the design of the experimental methods.

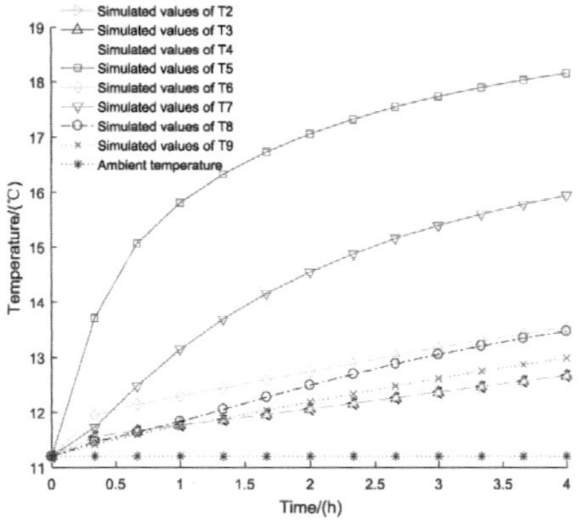

Figure 6. The simulated values of the feed system.

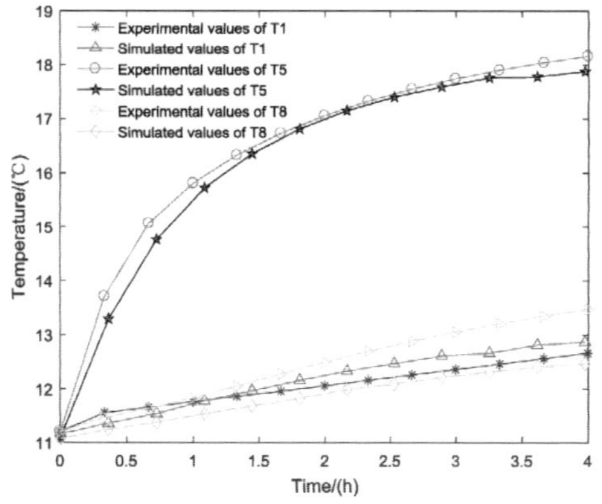

Figure 7 The comparison of the temperature values.

IOP Conf. Series: Materials Science and Engineering **234** (2017) 012013 doi:10.1088/1757-899X/234/1/012013

6. Experimental verification

To verify accuracy of the setting of "variable boundary conditions" and the method for loading of mobile heat source, the confirmatory experiment was carried out on the milling machine. The experimental site and test platform were shown in figure 8.

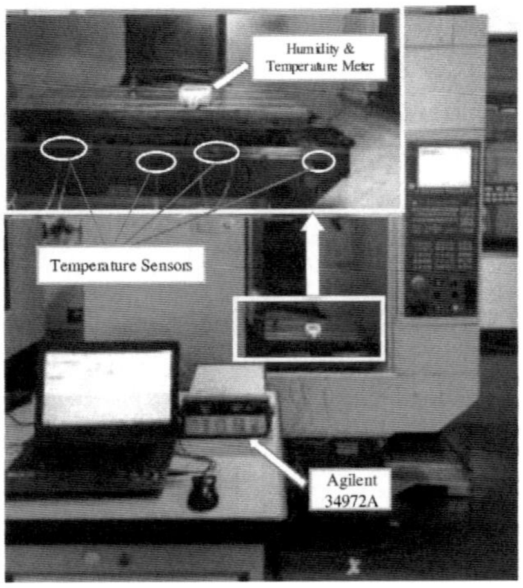

Figure 8 The test platform.

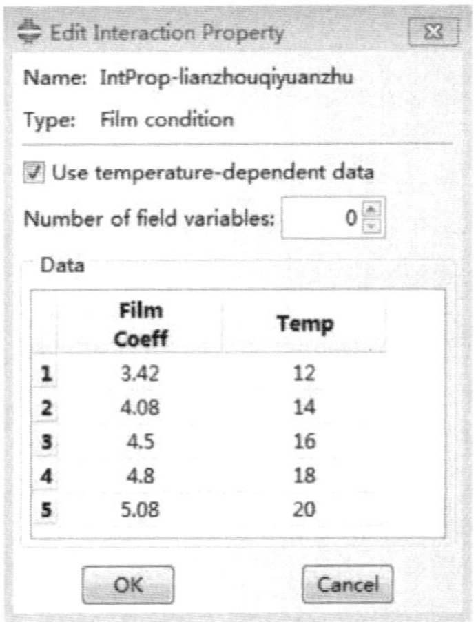

Figure 9. Variable CHCTs setting

The simulated results showed that the temperature of T5 was the highest, and the temperature difference between the four measuring points on the slider was very small. T7, T8 and T9 were fixed on a same part, so the temperature values of these were also very close. Finally, the simulated results of three measuring points which were fixed on the different components T1, T5 and T8 were selected to compare with the experimental values, as shown in figure 7. In addition, simulated and measured temperature values were also compared, which was shown in table 4. It could be seen that the simulation results were relatively close to the experimental results. Therefore, the loading of mobile heat source and boundary conditions of the FE model were correct.

Table 4. Comparison of the temperature values and simulation errors.

No. of the points	Simulated values($^\circ C$)	Experimental values ($^\circ C$)	Simulation errors (%)
T1	12.57	12.90	-0.026
T2	12.66	12.94	-0.022
T3	12.67	12.99	-0.025
T4	12.66	12.80	-0.011
T5	18.16	17.90	0.014
T6	13.54	13.06	0.035
T7	15.95	13.70	0.141
T8	13.48	12.51	0.072
T9	12.99	12.26	0.056

7. Conclusions

In this paper, a novel method of variable boundary conditions was proposed. According to the method, thermal stress, temperature field and thermal deformation of the feed system were simulated. In addition, the amount and time of the simulation were decreasing, and the validity and precision of the simulation results were confirmed by the experiment. It could provide a new idea for the setting of boundary conditions in the thermal analysis of NC machine tools, and it also has reference value and practical significance in the thermal analysis of the machine tools.

Acknowledgements
This work was supported by the High-End CNC Machine Tools and Basic Manufacturing Equipment Technology Major Projects (Grant no. 2013ZX04005-011).

References

[1] Won Soo Y, Kim Soo K and Dong Woo C 1999, International Journal of Machine Tools and Manufacture,39: 1087-1101.
[2] Shyh-Chour H. 1995, International Journal of Machine Tools and Manufacture, 35(8): 1099 - 1104.
[3] Jianfu Z, Pingfa F, Chen C, Dingwen Y, and Zhijun W. 2013, The International Journal of Advanced Manufacturing Technology, 68 (5-8): 1517-1527.
[4] Jungdo S and Daigil L. 2004, Composite Structures, 66 (1-4): 429-438.
[5] Xiuzhi Y, Shengfu Y, Rungang Y and Zimou T. 2010, J . Huazhong Univ . of Sci .& Tech . (Natural Science Edition), 38: 101-104.
[6] Zimou T, Baosheng Y and Xiuzhi Y. 2010, Hot Working Technology, 39:.130-137.
[7] Haitao Z, Jianguo Y and Jinhua S. 2007, International Journal of Machine Tools and Manufacture, 47 (6): 1003-1010.
[8] Dianxin L, Pingfa F, Jianfu Z, Zhijun W and Dingwen Y. 2014, The International Journal of Advanced Manufacturing Technology, 70: 1445-1454.
[9] Dai S. Application Manual of Machine Tool Rolling Bearings. China Machine Press, 1993.
[10] Min X, Shuyun J and Ying C. 2007, International Journal of Machine Tools and Manufacture, 47 (1): 53-62.

IOP Conf. Series: Materials Science and Engineering **234** (2017) 012014 doi:10.1088/1757-899X/234/1/012014

An Balancing Strategy Based on SOC for Lithium-Ion Battery Pack

Peng Li[1,a]

[1]Tongji Zhejiang College, Zhejiang,314006,China

[a]email:airrob@163.com

Abstract. According to the two kinds of working state of a battery pack, we designed a balancing strategy based on SOC, and expounds the working principle of balanced control strategy: the battery is charging, the battery charged state of the highest monomer battery is balanced discharge, strong single battery charging current decreases, while the other single cell in the same group is not affected; the battery is in a discharge or static state, single cell battery is the weakest balanced charge, while the other single cell in the same group are not affected. In this paper, we design a kind of lithium ion battery charging and discharging equalizer based on Buck chopper circuit and Boost-Buck chopper circuit. The equalizer is balanced charging and discharging experiments of series four lithium iron phosphate battery, the experimental results show that this equalizer has not only improved the degree not equilibrium between single cells, and improve the battery charge and discharge capacity.

1. Introduction

Due to the low nominal voltage of the single lithium ion battery, about 3.6V, we need to use a number of such cells in series to meet the different voltage requirements[1-4]. For single Lion battery, overcharge or over discharge will lead to reduced capacity, the impact of life, and even lead to direct battery damage or explosion[5-7].

Between each single battery in series in the process of using state of charge or voltage inconsistencies exist, such as the four monomer battery series shown in figure 1, in the charging process, if there is a single battery or reach full charging voltage, the charging process must stop; and in the process of discharge if a battery is empty or reach the discharge voltage, discharge must be stopped.

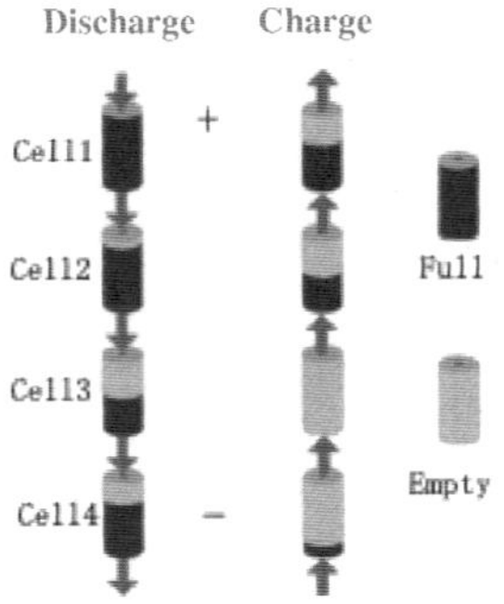

Figure 1. Charging and Discharging diagram of the serially connected battery cells

IOP Conf. Series: Materials Science and Engineering **234** (2017) 012014 doi:10.1088/1757-899X/234/1/012014

If you do not take measures to balance, tandem disproportion between single cells which increases along with the increase of the number of charge discharge cycles, the discharge capacity of battery charging will gradually decrease, resulting in battery scrap in advance, so we need to take effective measures to extend the balance of battery life, improve battery charge discharge capacity group.

2. Balancing strategies

According to the balancing of the single cell and the balancing energy flow, the balancing strategy can be divided into the following categories:

(1) for the battery in the battery terminal voltage or the state of charge of the highest battery balancing discharge, the energy is consumed by the resistor[8] or return to the battery pack [9]. This balancing strategy can only be used in the battery charging process, and the equilibrium energy is one-way, which can only be released by a balanced single cell.

(2) In the process of battery charging and discharging, equilibrium energy can only be transferred between adjacent monomer batteries[10,11,12]. This energy balance strategy for the balancing of the battery does not need to have a number of charging and discharging, on the one hand, affect the life of the battery, on the other hand, low speed, large switching loss.

(3) On the process of battery charging and discharging, the energy is transferred from the single cell in the middle of the battery to the lowest single cell based on the terminal voltage[13].

(4) in the process of battery discharge, the balancing strategy for the balancing of the single battery charge in the end of the battery pack[14].

(5) The equilibrium strategy used for battery discharge[15], balancing the energy through the transformer from the batteries to the group end single battery voltage minimum transfer; battery charging, energy balance from the batteries in the end single battery voltage the highest transfer to the battery pack by transformer. This equilibrium strategy is balanced by a single battery selectable and balanced energy is bidirectional.2.1 Analysis on Maglev electromagnetic environment.

In summary, from the analysis of the control ability of the equilibrium current, balancing scheme adopts inductance as storage elements, the control ability of the equilibrium current; from equilibrium strategy, equilibrium strategy of ideal is fifth, which is single cell equilibrium, and the equilibrium energy bidirectional flexible. But the fifth equilibrium strategies and equilibrium in single cell is balanced, the other cells in the same group are affected, and the completion of the first energy conversion, before and after the current single battery is balanced relatively large changes, affect the service life of the battery.

In the battery charging, the battery terminal voltage or state of charge of the highest monomer battery balancing discharge, energy balance from the strong single battery charging current equalizer to transfer the balance of the monomer battery decreases, and the same group of all other single cells is not affected, the the charging current is constant; when the battery charging, battery balancing of the weakest in the battery, energy balance by the equalizer to the weak single cell transfer, the discharge current balancing process of the weak single cells decreased, and the same group of all other single battery in the battery is not affected; static the can, with the above two kinds of equilibrium strategy, but in order to increase the battery energy storage, can be a choice of equilibrium strategy of battery discharge, so only the weakest in the battery cell is balanced charge, while the same group The current in other batteries is zero.

We have designed the balancer which has the following advantages:

1)The transformer is omitted, but also has the balancing of the monomer battery selectivity and balanced energy two-way;

2)Balanced current can be controlled, just adjust the duty cycle of the switch PWM, you can control the size of the current balance;

3)When the balance is only need to carry on the PWM control to a switch, other related switches need to be turned on only, so the switching loss is low, the equalizer control is simple, easy to realize.

3. Balancer Structure and Working Principle

3.1. alanced circuit topology

Figure 2 shows a battery pack equalization module that contains series single cells, which can be used to select the number of individual cells. The topology of the equalization circuit is shown in the dashed box in the figure2.

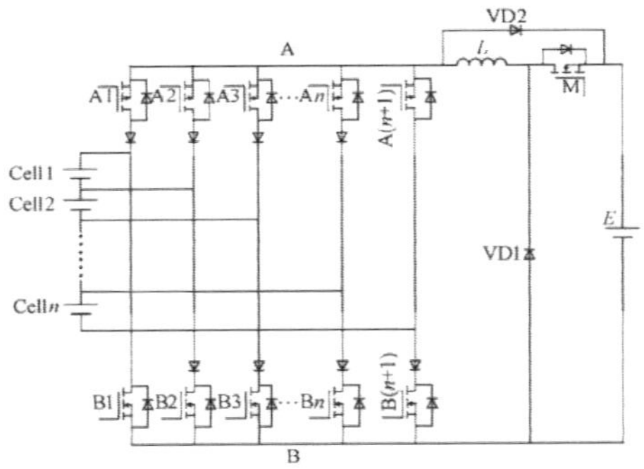

Figure 2. Balancing circuit topology

A bridge switch matrix (group A and group B), one inductor L, two freewheeling diodes VD1 and VD2, voltage source E and control power supply E power switch M. Due to the adoption of the group equalization, the utility model can select the switch device with small power, and the small power

Field effect transistor MOSFET saturation resistance in general m, and small driving power, so use a small power MOS transistors as the power switch M and A, B two group switch device. Due to the existence of parasitic diode in the MOS transistors, in order to prevent the short circuit of the battery, we need to use the diode in series. According to the actual situation select the appropriate capacity and voltage of the same type of battery series as the voltage source E.

3.2. Balanced discharge control strategy and working principle

In the battery charging process, the group of SOC value within the single cell equilibrium discharge is the highest, the balance of energy transfer to the equalizer, in other words, to improve the capacity of the battery charging method of charging current is reduced by high energy battery in the battery charging process. Specific control process and working principle:

Figure 3 shows the maximum SOC value of the battery Cell1, when the battery Cell1 discharge switch B1 and A2 are turned on, the Cell1 can be stored in the inductor L through the circuit①. When the Cell1 discharge switch is switched off, the equalizer can store the energy in the inductor through the loop②. In order to reduce the switching loss and simplify the switch control, the balance of the switch in the lower arm of the B1 PWM control, so that the upper arm switch A2 has been in the conduction state. Balanced circuit for Boost-Buck chopper circuit[16] typical, as shown in Figure3b, where U is the single battery voltage equalization circuit has two working modes: continuous current mode and discontinuous current mode; switch driving signal and the equilibrium current waveform as shown in Figure 3C, by adjusting the duty ratio can adjust the equilibrium current the size of the transfer of a cycle in the size of the energy balance.

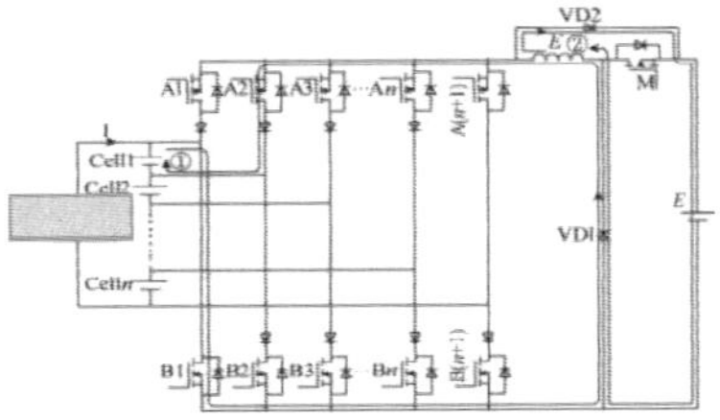

a. Balancing working principle

IOP Conf. Series: Materials Science and Engineering **234** (2017) 012014 doi:10.1088/1757-899X/234/1/012014

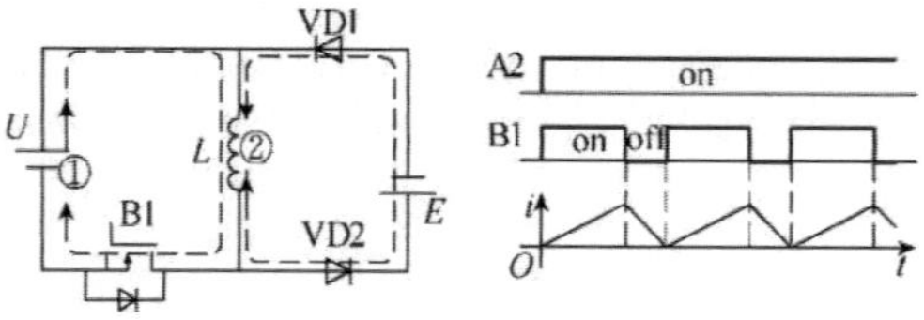

b. Boost-Buck equivalent circuit c. Waveform

Figure 3. Balancing working principle

3.3. Balanced charging control strategy and working principle

In the process of battery discharge or static, the single cell of the lowest SOC value in the group is balanced. The discharge current decreases low energy single battery in the battery discharge process to improve the discharge capacity of the batteries; while standing on the weak single cell energy supplement, improve the battery energy storage.

Balanced circuit for the typical Buck chopper circuit [16], the specific control process and working principle:Battery Cell2 hypothesis SOC in Figure 5 the minimum balance needed to charge A2 Cell2 switch and B3 has been in the conducting state, and the power switch M PWM control, when M conduction through the loop① to the equalizer inductor and the battery Cell2 release energy, when M can open off amount of inductor. After the loop② is transferred to Cell2.

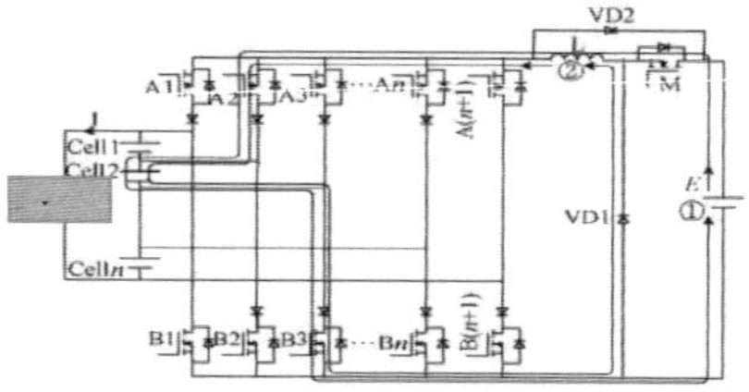

a. Balancing working principle

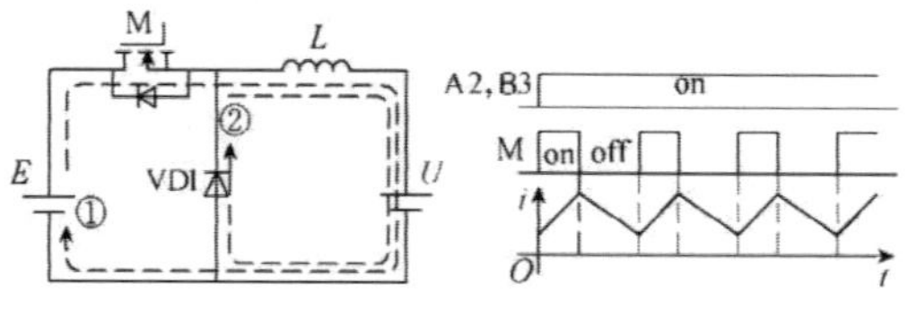

b.Buck equivalent circuit c.Waveform

Figure 4. Balancing working principle

As shown in Figure 4b, the equalizer circuit is a typical Buck chopper circuit. As shown in Figure 4C, A2 Cell2 in the process of equalization charging switch and B3 has been in a conducting state, only the PWM control of the switch M, the equilibrium current as shown in the graph, by adjusting the duty cycle to achieve equilibrium current can be controlled and adjusted.

3.4.Characteristic for Balancer and Balancing strategy

A single cell n contains the equalization module, and its monomer batteries are Cell1, Cell2, Cell3, Cellx, Celln, the equalization module has the following characteristics:

(1) The bridge matrix circuit contains a number of switches of 2(n+1).

(2) When the battery is charged, the strongest single battery in the battery pack is discharged to the equalizer, assuming that the battery Cellx is the strongest single cell, and only need to switch the Cellx and the A (x+1) of the two discharge switch of the battery Control, so that the upper arm switch A

(x+1) has been on, and the lower arm switch Bx PWM control can be achieved by the power of the battery to the equalizer Cellx transfer. the battery discharge or static, equalizer for weak single battery in the battery charging, the battery Cellx is single cell hypothesis of the weakest, only need to control the charging switch Ax and Cellx two B (x+1) has been conducted, and the power switch M PWM control can be achieved by the energy balance to the weak battery Cellx transfer.

4. Balance experiment

4.1. Battery Management System

The battery management system includes the functions of single cell parameter detection, balance management, thermal management, data processing and communication.

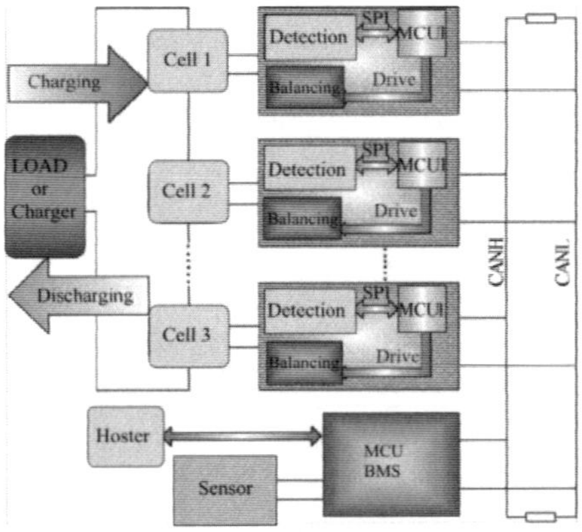

Figure 5. Battery management system

Balanced system is an integral part of the battery management system is an important part, as shown in figure 5 of the battery management system, the battery management and equalization. Each group contains a master controller to realize parameter detection and processing, and then according to the SOC value of each single cell, the corresponding switch action is driven in the equalization module. Therefore, the accurate estimation of SOC is the precondition to ensure the balanced experiment.

4.2. E-SOC curve extraction of LiFePO4 battery

This method adopts electric potential method and the combination of ampere hour method to estimate SOC value and EMF method needs E-SOC curve extraction of single cell. The E-SOC curve extraction experiment was carried out on the lithium iron phosphate battery with a rated capacity of 60A, h and rated voltage of 3.2V, using the c-KGCFS micro battery charging and discharging power supply. As shown in Figure 6, first of all, the SOC curve and discharge SOC curve of the lithium iron phosphate battery are extracted, and then the average value of the two is E-SOC curve. The measured E-SOC curves are stored in EEPROM for the controller to read at any time to correct the initial value of SOC.

4.3. Balance experiment

4.3.1. Equipment and Balancer circuit device. The experimental,which contains one hundred series of lithium iron phosphate battery and battery management system, rated capacity of single battery is 60A/h, rated voltage of 3.2V. Select the four single cells to complete the equalization experiment, the initial SOC value of 39.5%, respectively, 30%, 30% and 20%.

C-KGCFS, A battery charge and discharge power, the DC output voltage of 0 ~ 350V continuously adjustable DC output current of 0 ~ 500A continuously adjustable, can be realized on the battery constant current charging, constant voltage charging, constant current limiting static discharge, etc..

The 3 E cells with rated capacity of 20A, h and rated voltage of 3.2V are used as the power supply in the equalizer, and the initial SOC value of each cell is 60%.

The Balancer used in the experiment is based on the design of 6 lithium ion cell, in order to reduce the loss of the equalization circuit and choose a small power switching device, the specific selection is: A group switch using P channel MOS transistors SPD50P03L, the rated voltage and current were 30V And 50A, the maximum drain source resistance value of 12.5mmΩ; group B and M switch . The N channel MOS transistors IPD135N03L, the rated voltage and current are respectively 30V and 30A, the maximum drain source resistance value is 13.5m; equalization module in series diode and two diode using 42CTQ030 Schottky diode, the voltage and current ratings are 30V and 40A, the maximum turn-on voltage drop of 0.38V.

4.3.2. Charging experiment. The 10A current is supplied by the c-KGCFS power supply to charge the four series batteries, and when the SOC of the battery pack rises to 80%, the charging is stopped. The SOC curve of each cell in the equilibrium process is shown in figure 6, the charging time is 181min, the SOC value of each monomer battery after charging is 80%, 79%, 79% and 69.5%, respectively.

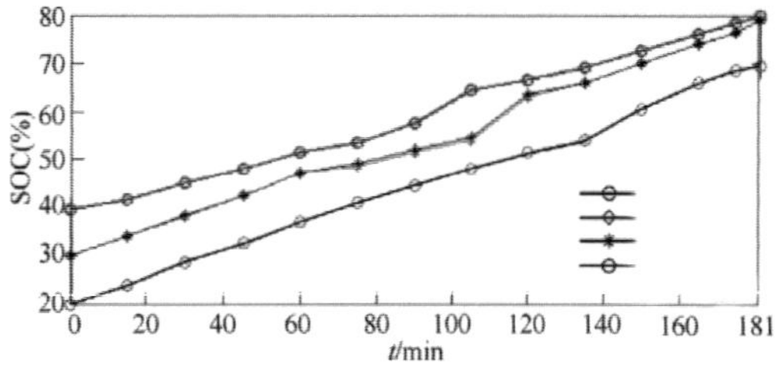

Figure 6.SOC curves in battery charging

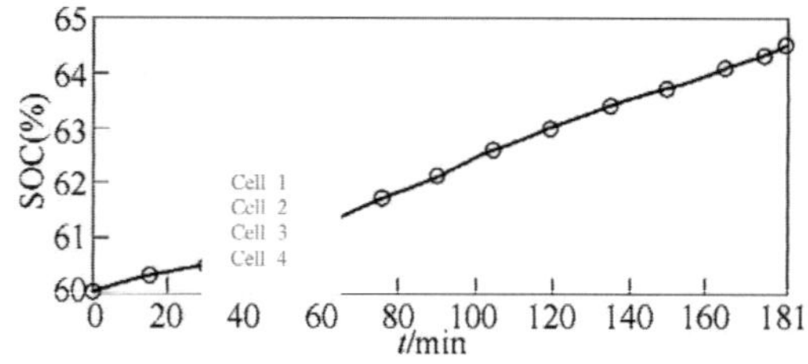

Figure 7. SOC curve of the power E

The SOC curve of the power supply E is shown in figure 7. At the end of the experiment, the SOC value of each individual cell is 64.5%.

4.3.3. Discharging experiment. The 10A current is supplied by the c-KGCFS power supply to discharge the four series batteries, and when the SOC value of the battery group is reduced to 20%, the discharge experiment is stopped.

The SOC curve of each cell in the equilibrium process is shown in figure8, the discharge time is 210 min, the SOC value of each monomer battery after discharge is 21%, 20.5%, 20.5% and 20%, respectively. The SOC curve of the power supply E is shown in figure 9, the SOC value before the experiment is 64.5, at the end of the experiment, the SOC value of each single cell as 52%.

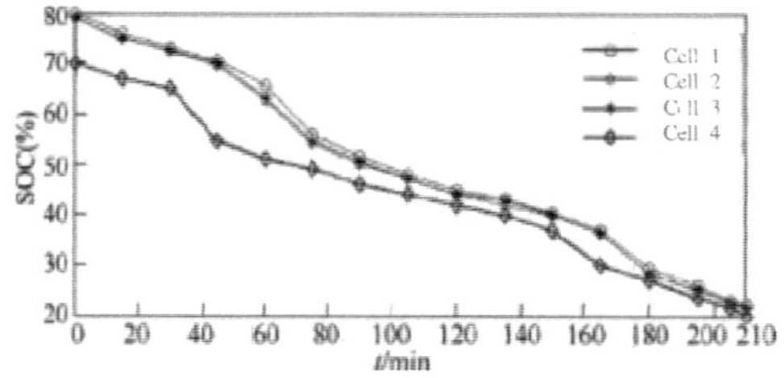

Figure 8. SOC curves in battery discharging

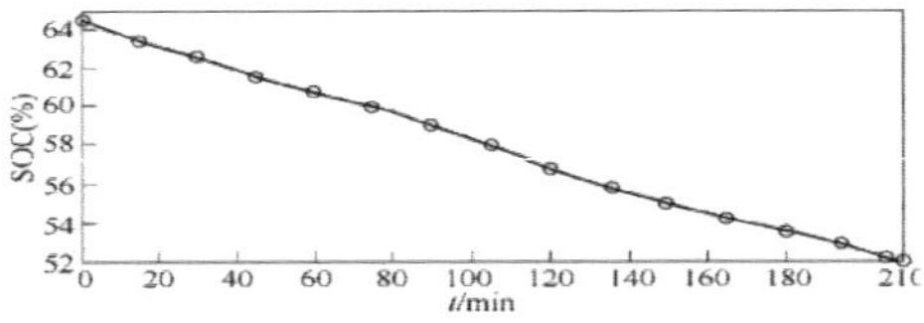

Figure 9. SOC curve of the power E

4.4. Results analysis

The initial SOC four single cell values were 39.5%, 30%, 30% and 20%, if you do not take measures to balance in the charging process, so when the battery SOC value rose to 80% when the single battery SOC values were 80%, 70.5%, 70.5% and 60.5%, but as shown in Figure 7 because, to take a balanced discharge on the No. 1 battery in the charging process, therefore its SOC curve rise speed decreased, thus charging after the expiry of the time delay, charging at the end of each monomer.The SOC values of the cells were 80%, 79.5%, 79.5% and 69.5%.

In charge of the experiment, the balanced energy transfer from 1 batteries to power E, changes in the charge value of SOC before and after the experiment can calculate the ampere hour capacity 1 batteries for 5.1A h by the release of No. 1 batteries and 2 batteries, and can calculate the power ampere hour capacity of E absorption was 2.7A h by SOC curve in Figure 8, which can balance the energy transfer efficiency was calculated as 53% battery charging process. In the experiment, the equilibrium energy is transferred from the power supply E to the 4 cell, so the equilibrium energy transfer efficiency of the battery during discharge is calculated according to the experimental results, which is about 68%. It can be seen that the equilibrium energy transfer efficiency is relatively high, and the loss of equalizer is mainly the loss of MOS transistors and diode. The soft switching technique can be used to further improve the efficiency of energy transfer.

5.Conclusion

Equalizer to inductance is the energy storing element, take a balanced strategy of the strong single battery battery in the battery, and the battery discharge and take static equilibrium strategy to charge the battery in a single cell. The equalization current is controllable and the energy transfer efficiency is high, and the equalization circuit is simple and easy to realize. The charging and discharging experiments of four series lithium iron phosphate batteries show that the equalizer improves the charge capacity and discharge capacity of the whole battery pack.

In order to reduce the switching loss and select the rated voltage and rated current of the switch device is relatively low, so the battery pack can not be too much in a single cell battery. Power supply of the E equalizer is composed of each monomer battery additional series, need to provide management for the single battery, in actual use, according to the number of equalizer in the actual system, the battery will all equalizer of grouping management. The equalizer can be used in the balance management of lithium ion batteries and the balance management of energy storage system.

Acknowledgments

This research was supported by Educational Commission of Zhejiang Province, China.
Project number: Y201534868

References

[1] Hsieh Y C, Wu J L, Chen X H. Class-E-based charge-equalization circuit for battery cells[J]. IET Power Electronics, 2012, 5(7): 978-983.

[2] Cheng K W E, Divakar B P, Wu Hongjie,Battery-Management system(BMS) and SOC development for electric vehicles[J]. IEEE Transactions on Vehicular Technology,2011,60(1): 76-88.

[3] Wang Le Yi, Polis Michael P, Yin G George, et al. Battery cell identification and SOC estimationusing string terminal voltage measurements[J]. IEEE Transac-tions on Vehicular Technology, 2012, 61(7): 2925-2935.
[4] Einhorn Markus, Guertlschimd Wolfgang, Blochberger Thomas. A current equalization method for serially connected battery cells using a single power converter for each cell[J]. IEEE Transactions on Vehicular Technology, 2011, 60(9): 4227-4237.
[5] Einhorn Markus, Roessler Werner, Fleig Juergen. Improved performance of serially connected Li-Ion batteries with active cell balancing in electric vehicles [J]. IEEE Transactions on Vehicular Technology, 2011, 60(6): 2448-2457.
[6] Gao K Zhi, Bo Z Chun, Gui L Ren, et al. Comparison and evaluation of charger equalization technique for series connected batteries[C]. 37th IEEE Power Elec-tronics Specialists Conference, 2006: 1-6.
[7] Luo Yutao, Zhang Zhiming, Zhao Keqiang. A novel distributed equilibrium and management system of dynamic battery pack[J]. Transactions of China 2008, 23(8): 131-136.
[8] Cao Jian, Schofield Nigel, Emadi Ali. Battery balancing methods: a comprehensive review[C]. IEEE Vehicle Power and Propulsion Conference, 2008:1-6.
[9] Kimball J W, Krein P T. Analysis and design of 20th Annual IEEE Applied Power Electronics Conference and Exposition, 2005, 3:1473-1477.
[10] Baughman Andrew, Ferdowsi Mehdi. Analysis of the double-tiered three-battery switched capacitor battery balancing system[J]. 2006 IEEE Vehicle Power and control Propulsion Conference, 2006:1-6.
[11] Wang Liye, Wang Lifang, Liao Chenglin. Based one nergy transferring for battery packs applied on electrical vehicle[C]. 2010 International Conference and Communication Technologies in Agriculture Engineering, 2010: 271-274.
[12] Liu Hongrui, Xia Chaoying. An active equalizer for serially connected lithium-ion battery cells[C]. 2nd International Conference on Energy and Environmental Protection, 2013: 809-812.
[13] Liu Hongrui, Xia Chaoying. Discussion on a new battery equalization method for electric vehicle[J]. Automotive Engineering, 2013, 35(10): 934-938.
[14] Imtiaz A M, Khan F H. Time shared flyback converter based regenerative cell balancing technique for series connected Li-Ion battery strings[J]. IEEE Transaction on Power Electronics, 2013, 28, (12): 5960-5975.
[15] Lee Yuang-Shung, Cheng Mingwang. Intelligent battery strings[J]. IEEE Transactions on Idustrial Electronics, 2005, 52(5): 1297-1307.
[16] Wang Zhaoan, Huang Jun. Power electronics technology [M]. Beijing: China Machine Press.

IOP Conf. Series: Materials Science and Engineering **234** (2017) 012015 doi:10.1088/1757-899X/234/1/012015

The Formation of Vortex Structures in a Screen Cylinder Wake

Azlin Mohd Azmi[1], Yucen Lu[2] and Tongming Zhou*[2]

[1]Faculty of Mechanical Engineering, Universiti Teknologi MARA, 40450 Shah Alam, Selangor, Malaysia
[2]School of Civil, Environmental and Mining Engineering, The University of Western Australia, 35 Stirling Highway, Crawley, WA 6009, Australia

*Corresponding Author: tongming.zhou@uwa.edu.au

Abstract. The formation of vortex structures in a screen cylinder wake was investigated in a wind tunnel at a Reynolds number of 7000. The screen cylinder was made of a stainless steel wire mesh with an open area ratio of 67%. The results showed that the screen cylinder wake could be classified into two distinct regions. The first region was characterised by the development of the shear layer vortices which resulted from Kelvin-Helmholtz instability. At about $x/d = 20$ (where d is the diameter of the cylinder) the shear layer vortices started to interact with each other across the centreline, and evolved downstream to form the alternately arranged 'large-scale' coherent structures. These structures were most pronounced at $x/d = 40$. The vortex formation region was therefore extended significantly downstream compared with that of the solid cylinder wake. The second region involved a gradual decay of the fully-formed large-scale structures, evidenced by the weak vorticity exchange across the wake centreline.

1. Introduction

Fluid flow past a circular cylinder may induce vibration of the structure, which is normally known as the vortex-induced vibrations (VIV). VIV of cylindrical structures should be avoided as it will lead to fatigue damage of the structures. VIV can be suppressed using passive methods. Shroud is one of the passive devices which are found effective in suppressing VIV [1] (Zdravkovich 1981). Zdravkovich and Volk [2] (1972) conducted a test for three different shroud types; the square-holed, circular-holed and a fine-mesh gauze, all having 36% porosity, where the porosity is defined as the percentage of the open area over the total area. They found that the shroud made of fine-mesh gauze was the most effective in suppressing vibration of the cylinder, giving the highest reduction in the double-amplitude of vibration. The gauze shroud affected the pressure distribution more significantly than the other shrouds for which there was a radical departure in the pressure distribution over the rear portion of the cylinder (between 80° and 160°). In a recent study by Oruç [3] (2012), a circular cylinder wake was controlled by using a screen with a porosity of 50% formed in a streamlined geometry enclosing the cylinder. The PIV results over the region of $x/d \leq 4$ showed that the existence of the control screen around the cylinder significantly suppressed the interaction of the shear layers and hence the vortex shedding from the surface of the cylinder, resulting in a significant reduction in turbulence intensity, Reynolds shear stress and turbulent kinetic energy. Ozkan et al. [4] (2012) found that a better flow control could be achieved with their screen shroud with an outer-to-inner diameter ratio of $1.6 - 2.0$

and porosity of $40-60\%$. While the existing studies proved that the perforated shrouds are effective in suppressing VIV of a circular cylinder, this type of suppression device is largely unexplored, especially the one made of a screen gauze. In the present study, the primary aim is to explore the vortex formation and development in a screen cylinder wake by examining the evolution of its coherent structures. The second aim is to propose a conceptual wake model in its wake, illustrating the formation process and the momentum transport behaviour. The above results will be compared with wakes generated by a solid cylinder and a screen strip. To our knowledge, this is the first attempt to quantify the large-scale structures in the near and intermediate regions of a screen cylinder wake.

2. Experimental Details

The experiments were conducted in a blower type wind tunnel with a test section of 380 mm (width) × 255 mm (height) and 1.8m (long). The free stream velocity was uniform to 0.2% and the longitudinal turbulence intensity was less than 0.5%. Two types of cylinders were used, namely, a solid cylinder and a screen cylinder. The diameters d of the solid and the screen cylinders were 10 mm and 21 mm, respectively. The length l of both cylinders was 380mm, giving an aspect ratio l/d of 38 for the solid cylinder and 18.1 for the screen cylinder. All measurements were performed at Re = 7000, corresponding to a free stream velocity U_∞ of 10.4 m/s and 5 m/s for the solid and screen cylinders, respectively. Two end plates were used for both cylinders. The thickness of the plates is 1.25mm with the leading and trailing edges being sharpened at an angle of 45°. They were fixed at about 4mm from the tunnel side walls to keep them above the boundary layer. The screen cylinder was made of a stainless steel wire mesh of aperture a = 2 mm and wire diameter dw = 0.45 mm. The porosity of the screen mesh β was 67%. Care was taken to make sure that its diameter was uniform along the cylinder axis (with a tolerance of ±1%). Figure 1 shows the definition of the co-ordinate system, wake profile and the photo of the screen cylinder. The measurement locations were at $x/d = 10-60$. To avoid crowding the figures, only results at selected locations are shown. To examine the streamwise evolution of the wakes, an X hot-wire probe was moved across the wake in the y-direction to measure the longitudinal and transverse velocity components, u and v, respectively. Another X-probe was fixed at the wake edge and at the same streamwise location as the first X-probe to provide a phase reference to the measured velocity signals for phase-averaged analysis. The separation between the two wires of the X-probes was about 1.5mm. The hot wires were 5 μm in diameter and 1 mm in length. The output signals from the anemometers were low-pass filtered at a cut-off frequency f_c of 2800 Hz and sampled at a frequency f_s of 5600 Hz using a 16-bit A/D converter. The sampling period T_s was 30s.

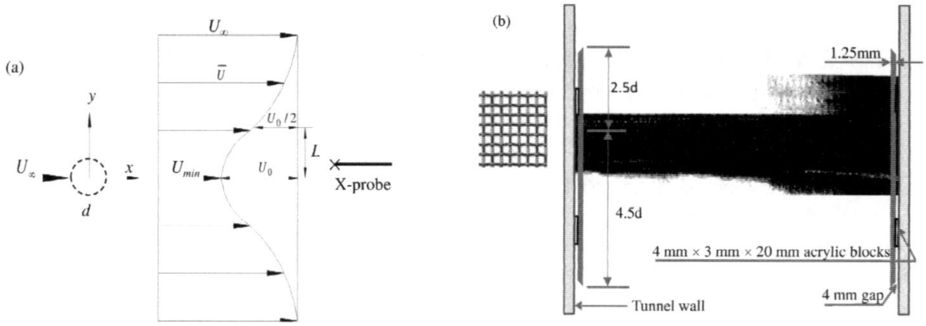

Figure 1. Co-ordinate system, wake profile and probe arrangement (a) and the top view of the arrangement of the screen cylinder (b). Please note that the figures are not drawn to scale.

3. Results and Discussion

3.1. Energy spectra in the two wakes

The spectra ϕ_v for both cylinders at various downstream locations are shown in Fig. 2. Both spectra are obtained at $y^* = 0.5$ in the two wakes (Figs. 2a and 2b) by calculating the Fast Fourier Transform (FFT) of the velocity signals. The normalized spectral density function is defined such that $\int_0^{\infty} \phi_v\ (x)dx = \langle 1 \rangle$, where x represents the frequency, which is normalized by U_∞ and d i.e. $x = f^* = fd/U_\infty$. With this normalization, the peak frequency on the spectrum corresponds to the Strouhal number. Each spectrum has been shifted downward by one order relative to the one above it for easy viewing. For the solid cylinder wake (Fig. 2a), vortex shedding at $x^* = 10$ is apparent at a single frequency f of 201 Hz, which corresponds to $f^* = 0.2$ as indicated by the sharp peak. This peak location is consistent with previous literatures at subcritical Reynolds numbers. The ratio between the peak height and the plateau region reduces with x^* although f remains the same. This trend indicates that the vortices are decaying in the streamwise direction. On the other hand, the energy spectra of the screen cylinder wake reveal no significant peak at location immediately downstream of the cylinder ($x^* = 10$). This result suggests that vortex shedding is not apparent, which highlights the difference in the near-wake behaviour of both cylinders. A broad-band peak at $f^* = 0.277$ starts to emerge at $x^* = 20$, indicating the emergence of some periodicity of the organised fluid structures in the wake. The peak at $f^* = 0.257$ becomes apparent at $x^* = 30$, indicating the formation of the large-scale structures. Further downstream, the broad-band peak is still apparent with a slightly decreasing frequency trend. For $x^* = 30{\sim}60$, the peak frequency decreases from 0.257 to 0.234. On the other hand, the spectra along the outer edge (Fig. 2c) display broad-band peaks even in the near wake, indicating the existence of periodicity at all measured downstream locations. Huang and Keffer [5] (1996) reported the same behaviour for their screen strip wake ($\beta = 40\%$ and $Re = 1.1 \times 10^4$) at $x^* = 1$ and attributed this to the Kelvin-Helmholtz instability of vortex sheets generated at the mesh edge and the subsequent rolled up of the sheets into discrete vortices, similar to the single shear layer development. The periodicity becomes more apparent and the peak magnitude becomes larger from $x^* = 10 - 30$. The decrease of the broad peak location towards lower frequencies along the streamwise direction indicates the growth of the vortical structures within the shear layers. It is hypothesized that this behaviour is due to an increase in the average wavelength, assuming slowly increasing convective velocity. The above process may imply merging of these vortices, which accounts for the formation of the large-scale structures in the screen cylinder wake, similar to the findings of Huang and Keffer [5] (1996) and Antonia and Mi [6] (1998). The continuous decrease of the peak frequency indicates that the merging process does not occur at a fixed downstream location.

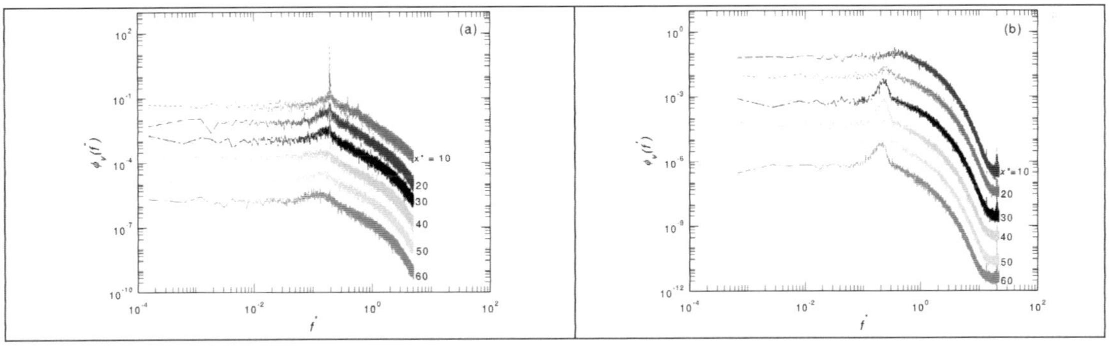

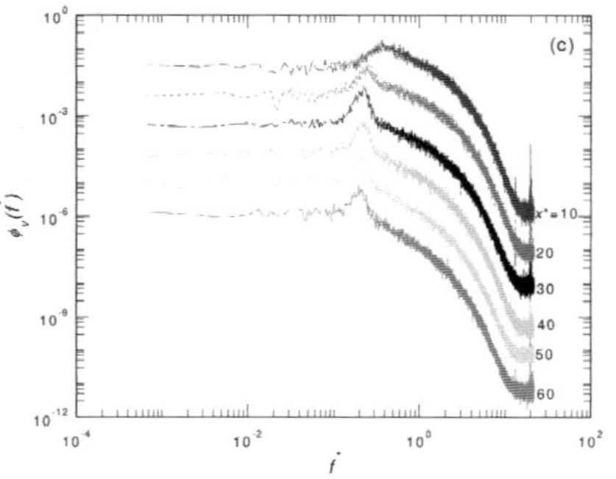

Figure 2. Energy spectra $\phi_v(f^*)$ in the wake of (a) solid cylinder and (b) screen cylinder measured at $y^* = 0.5$ and (c) screen cylinder measured at the wake edge.

3.2. Phase-averaged coherent vorticity field

A turbulent signal B can be decomposed into a mean value, $\overline{B}$, and a fluctuation component β which can further be decomposed into a coherent component $\tilde{\beta}$ and the reminder component β_r ([7] Hussain and Reynolds 1970), i.e.

$$B = \bar{B} + \beta = \bar{B} + (\tilde{\beta} + \beta_r). \quad (1)$$

Phase average method allows the coherent structures at a given frequency to be separated. Details of the phase-averaged methods can be found in Zhou et al. [8] (2003). After the detection of the coherent signals $\tilde{u}$ and $\tilde{v}$, the phase-averaged vorticity $\tilde{\omega}_z$ is obtained using the following relationship

$$\tilde{\omega}_z = \frac{\partial(\bar{V}+\tilde{v})}{\partial x} - \frac{\partial(\bar{U}+\tilde{u})}{\partial y} \approx \frac{\Delta\tilde{v}}{\Delta x} - \frac{\Delta(\bar{U}+\tilde{u})}{\Delta y}, \quad (9)$$

where $\Delta x = -U_c\Delta t = -U_c/f_s$ is the separation in the streamwise direction and can be obtained using Taylor's hypothesis. The average convection velocity U_c of the vortices is given by the velocity $\overline{U} + \tilde{u}$ at the vortex centre, which is identified with the location of the maximum phase-averaged vorticity, $\tilde{\omega}_{max}$. Meanwhile Δy is the distance between two adjacent measurement locations across the wake.

The iso-contours for the phased-averaged vorticity of the solid cylinder and the screen cylinder wakes are shown in Figs. 3 and 4, respectively. The errors in measuring the vorticity components in both wakes depend on that for velocity component measurements and distance measurements, which are estimated at about 13%. Due to regular shedding of the vortices in the wakes, a shedding period can be evaluated, which corresponds to a phase angle from 0 to 2π. Over the measurement period (30 s) at each y location, about 2000 periods of data for both the screen and the solid cylinders were obtained. To use the phase averaged method, we divided one period into 30 phases and then average the measured data at a given phase over the 2000 periods (i.e. corresponding phase angle from 0 to 2π). In the figures showing the contours, we actually plotted also the phase angles over the range -2π – 0 by mirroring those from 0–2π to show clearer pictures of the flow structures. The phase ϕ, ranging from -2π to +2π, can be interpreted in terms of a streamwise distance and ϕ=2π corresponds to the vortex wavelength λ (=Uc/f). The flow direction is from left to right. The positions of the foci and the saddle points associated with the vortices are determined from the sectional streamlines (Figures are not shown here) and denoted by '+' and '×', respectively. The diverging separatrices are shown as dashed lines passing through the saddle points. The diverging separatrices, i.e. the braids, connects the top of a structure to the bottom of an adjoining structure as described in Hussain and Hayakawa [9] (1987). In general, the foci correspond quite well with the maximum vorticity. There are significant variations in terms of the evolution of the shape, strength and size of the vortices in the two wakes. The phase-averaged coherent vorticity contours in the solid cylinder wake (Fig. 3) clearly depict the large-scale vortical structures and their evolution in the streamwise direction, consistent with Zhou et al. [8] (2003), verifying the experimental result of the solid cylinder wake. For example, at $x^* = 10$, the $\widetilde{\omega}_z^*$ contours display the well-known Kármán vortex street. The maximum concentration for $\widetilde{\omega}_z^*$ is 1.1 occurring at $y^* \cong 0.3$. The positive and negative vortices penetrate across the wake centreline to the other side of the wake, showing a staggered vortex pattern. The $\widetilde{\omega}_z^*$ contours at $x^* = 20$ display similar vortex structures but with a lower maximum concentration of 0.4. This maximum concentration decreases to 0.12 at $x^* = 40$, reflecting an apparent reduction in vortex strength as x^* increases. Furthermore, at $x^* = 40$, the staggered pattern of the alternating counter-rotating vortices starts to diminish. For the screen cylinder wake (Fig. 4), the contours of $\widetilde{\omega}_z^*$ at $x^* < 30$ are less organized. There are small-scale vortices in the two shear layers, separated by a core layer at the central region. The interaction between the opposite signed vortices starts at $x^* = 20$, where the core layer starts to undulate. At $x^* = 30$, the spanwise vorticity contours become more apparent than that in the region closer to the cylinder, implying an amalgamation process at $x^* < 30$. Vortices still grow thereafter until $x^* = 40$, where the large-scale structures can be seen to be fully formed. Considering the coherent vorticity contours at $x^* \geq 40$, the above results suggest that the decay rate of the fully formed large-scale vortices in the screen cylinder wake is much smaller than that of the large-scale vortices in the solid cylinder wake. For instance, the maximum vorticity concentration decays by about 20% from $x^* = 40$ to 50 and 25% from $x^* = 50$ to 60 in the screen cylinder wake whereas a large decay of 64% is evident only from $x^* = 10$ to 20 in the solid cylinder wake. It can be surmised that due to this slow decay, vortical structures in the screen cylinder wake will persist a long distance downstream. It is interesting to note that at all downstream locations of the screen cylinder wake, there are contours enveloping the vortex cores on each side of the wake and a small amount of vorticity is also found between vortex cores, a feature that is absent in the solid cylinder wake. Huang et al. [10] (1996) also noticed this feature in the screen strip wake on the coherent vorticity contours through pattern-recognition analysis. They noticed that farther downstream at $x^* = 50$ and 100, however, the envelopes were absent and the periodicity was less prominent. The current work indicates that the envelopes are still present at $x^* = 50$ and $x^* = 60$, consistent with the discussion on the persistence of structures downstream in the screen cylinder wake.

4. Conclusions

The formation of the vortical structures in the near and intermediate regions ($10 \leq x^* \leq 60$) of a screen cylinder (with 67% porosity) wake are studied in a wind tunnel using a phase-averaged analysis at a Reynolds number of about 7000. Significant differences about the vortex formation, streamwise evolution and momentum transport are found as compared with that of a solid cylinder wake. The vortex formation region in the screen cylinder wake has been extended significantly. In the near-wake, small-scale vortices are formed in the shear layers due to Kelvin-Helmholtz instability. When evolving downstream, these shear layer vortices merge, grow in size and decrease in frequency. For about $x^* \geq 20$, the size of the vortices in the shear layers are large enough and they interact with each other across the centreline, and evolving downstream to form the large-scale coherent structures. These structures are most pronounced at $x^* = 40$ indicating the completion of the vortex formation process, where vortices are fully formed. After the complete formation of the large-scale structures, vortices decay at a much slower rate in a screen cylinder wake compared with that in the solid cylinder wake, a decay by only about 20% from $x^* = 40$ to 50 in the former, in comparison to a decay of about 61% from $x^* = 10$ to 20 in the latter, which may imply the longer persistence of the large-scale structures in the screen cylinder wake. The transport of vorticity shows that vortices in the solid cylinder wake interact vigorously with those across the centreline, which greatly accelerates its decay, while in the screen cylinder wake, the vorticity exchange across the adjacent vortex border is weak where their vorticity vectors mostly occur along the diverging separatrices, accounting for the slow decay in the streamwise direction. The above results support the two regions of structural evolution in the screen cylinder wake which involve the growth of the shear layer vortices and the slow decay of the fully-formed 'large-scale' vortices in comparison to the rapid decay of coherent vortices throughout the solid cylinder wake.

References

[1] Zdravkovich M M (1981) Review and classification of various aerodynamic and hydrodynamic means for suppressing vortex shedding. *J Wind Eng Ind Aerod* **7** 145
[2] Zdravkovich M M, Volk J R 1972 Effect of shroud geometry on pressure distributed around a circular cylinder. *J Sound Vib* **20** 451
[3] Oruç V (2012) Passive control of flow structures around a circular cylinder by using screen. *J Fluid Struct* **33** 229
[4] Ozkan G M, Oruc V, Akilli H, Sahin B 2012 Flow around a cylinder surrounded by a permeable cylinder in shallow water. *Exp Fluids* **53**1751
[5] Huang Z, Keffer J F 1996 Development of structure within the turbulent wake of a porous body. 1. The initial formation region. *J Fluid Mech* **329**103
[6] Antonia R A, Mi J 1998 Approach towards self-preservation of turbulent cylinder and screen wakes. *Exp Therm Fluid Sci* **17** 277
[7] Hussain A K M F, Reynolds W C 1970 The mechanics of an organized wave in turbulent shear flow. *J Fluid Mech* **41** 241
[8] Zhou T, Zhou Y, Yiu M W, Chua L P 2003 Three-dimensional vorticity in a turbulent cylinder wake. *Exp Fluids* **35** 459
[9] Hussain A K M F, Hayakawa M 1987 Eduction of large-scale organized structures in a turbulent plane wake. *J Fluid Mech* **180** 193
[10] Huang Z, Kawall J G, Keffer J F 1996 Development of structure within the turbulent wake of a porous body. 2. Evolution of the three-dimensional features. *J Fluid Mech* **329**117

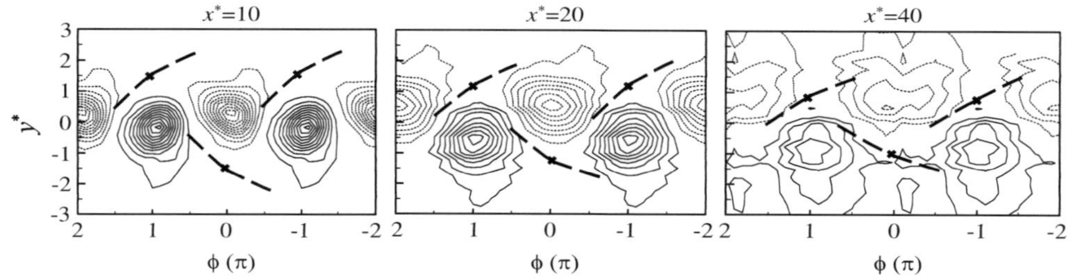

Figure 3. Phase-averaged coherent spanwise vorticity, $\widetilde{\omega}_z^*$, for the solid cylinder wake at (a) $x^* = 10$ (minimum contour value -1.0, maximum contour value 1.1, contour interval 0.1); (b) $x^* = 20$ (-0.35,0.4,0.05) and (c) $x^* = 40$ (-0.12,0.12,0.03) . Vortex centers and saddles are marked by a plus and a cross, respectively. The thick dashed lines represent the diverging separatrices which pass through the saddles.

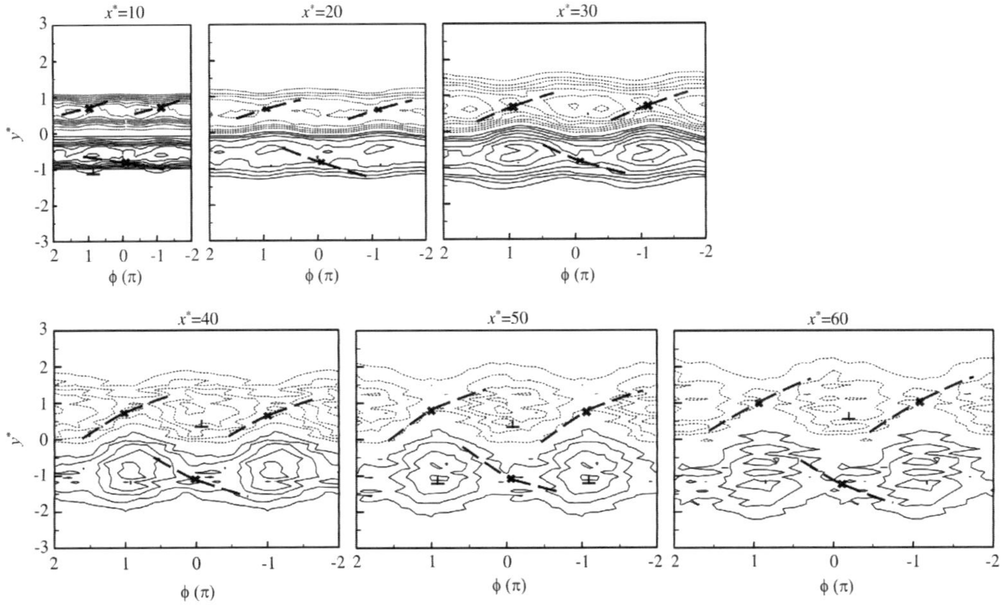

Figure 4. Phase-averaged coherent vorticity, $\widetilde{\omega}_z^*$, for the screen cylinder wake at (a) $x^* = 10$ (-0.8,0.8,0.1); (b) $x^* = 20$ (-0.5,0.5,0.1); (c) $x^* = 30$ (-0.35,0.35,0.05); (d) $x^* = 40$ (-0.25,0.25,0.05); (e) $x^* = 50$ (-0.2,0.2,0.05) and (f) $x^* = 60$ (-0.15,0.15,0.05). Vortex centers and saddles are marked by a plus and a cross, respectively. The thick dashed lines represent the diverging separatrices which pass through the saddles.

IOP Conf. Series: Materials Science and Engineering **234** (2017) 012016 doi:10.1088/1757-899X/234/1/012016

Effect of Free-Stream Turbulence Intensity on Transonic Airfoil with Shock Wave

I Lutsenko, M Serikbay, A Akiltayev, L R Rojas-Solórzano and Y Zhao

Department of Mechanical Engineering, School of Engineering, Nazarbayev University, Astana 010000, Republic of Kazakhstan

ilya.lutsenko@nu.edu.kz, manarbek.serikbay@nu.edu.kz, ali.akiltayev@nu.edu.kz

Abstract. Airplanes regularly operate switching between various flight modes such as take-off, climb, cruise, descend and landing. During these flight conditions the free-stream approaching the wings undergo fundamental changes. In transonic flow conditions, typically in the military or aerospace applications, existence of nonlinear and unsteady effects of the airflow stream significantly alters the performance of an airfoil. This paper presents the influence of free-stream turbulence intensity on transonic flow over an airfoil in the presence of a weak shock wave. In particular, NACA 0012 airfoil performance at Ma_∞=0.7 is considered in terms of drag, lift, turbulence kinetic energy, and turbulence eddy dissipation parameters under the influence of varying angle of attacks and free-stream turbulence. The finite volume method in a commercial CFD package ANSYS-CFX is used to perform the numerical analysis of the flow. Mesh refinement using a mesh-adaption technique based on velocity gradient is presented for more accurate prediction of shocks and boundary layers. A Shear Stress Transport (SST) turbulence model is validated against experimental data available in the literature. Numerical simulations were performed, with free stream turbulence intensity ranging from low (1%), medium (5%) to high (10%) levels. Results revealed that drag and lift coefficients are approximately the same at every aforementioned value of turbulence intensity. However, turbulence kinetic energy and eddy dissipation contours vary as turbulence intensity changes, but their changes are disproportionally small, compared with values adopted for free-stream turbulence.

1. Introduction

Nowadays one of the most complex problems under discussion in various engineering applications, such as aerospace, wind turbines and buildings is that of turbulence. By definition, it is a flow regime characterized by unpredictable unsteadiness and multidimensional changes in fluid flow properties around their time-averaged state. Along with its complex nature, it has significant effects on performance characteristics of internal and external flows, for example, the flows around an airfoil of a given airplane wing and a turbo machinery blade.

In the range of transonic speeds, the presence of free stream turbulence adversely affects the aerodynamic performance of an airfoil. Namely, it can affect boundary-layer transition, the development of turbulent boundary layers, separations and shock wave interactions [1]. These problems, in turn, cause alterations in pressure and skin drag force distributions over the surface of the airfoil. Hence, it is important to determine the effects of free-stream turbulence on the performance of a given airfoil in order to better control the lift and drag forces on the wing or blades and therefore increase, if possible, overall efficiency of the associated machinery operating with the airfoil in transonic flow. This investigation focuses on determining any potential relationship between turbulence intensity and lift-drag coefficients for a range of different angles of attack in transonic flow. Moreover, especial attention is paid to the effect of free stream turbulence intensity on the turbulence kinetic energy and dissipation around the airfoil.

Transonic flow is a flow regime in which the fluid transits between subsonic-sonic-supersonic conditions, where the entire regions on the surface are considered as mixed flow in which the local Mach number is either less or more than 1 with imbedded intermediate sonic pockets (see Figure 1) The transition from subsonic to supersonic condition is generally accompanied by the appearance of shock waves of infinitesimal thickness separating well differentiated regions of the flow. This type of airflow is commonly found in applications such as airplane wings and turbo machinery blades, where turbulence intensities range from a low level of 0.4% to high of 10% [2].

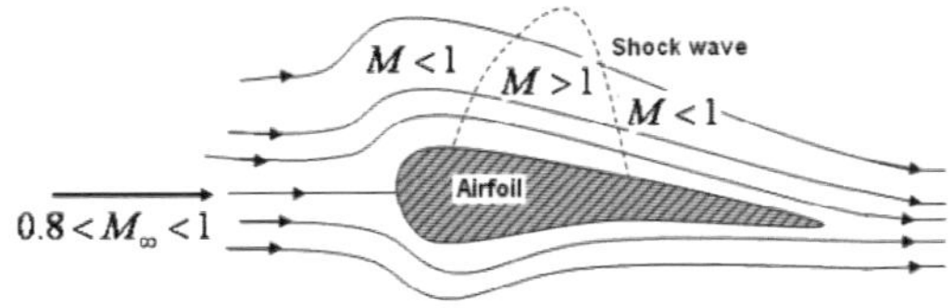

Figure 1. Transonic flow illustration.

Key parameters in this investigation are angle of attack (AOA) and levels of turbulence intensity (low, medium and high). The main purpose was to investigate if there could be significant effects of these parameters onto time-averaged dynamic phenomena such as lift and drag coefficients. In addition, turbulence eddy dissipation and kinetic energy were expected to vary in a non-linear manner.

2. Previous work

[2] studied the effect of turbulence intensity in the range of 2.4% to 5.4% on aerodynamic performance of airfoil NACA 0012 and found that the increase in air free stream turbulence intensity causes delay of the stall angle and the maximum lift coefficient. However, their results show that increase in intensity also causes increase in drag coefficient. Our study extends the study [2] by taking wider range of turbulence intensities, i.e., very low(1%), medium(5%) and very high(10%) and aims to observe whether this phenomenon occurs for such intensities.

Yap et al. [1] have shown that the NACA 0012 airfoil has a stall angle of 16 degrees and further increase of the angle leads to dramatic fall in lift due to rise in static pressure on the upper surface and significant increase in drag. Turbulence modeling for unsteady transonic flows over the rear half of an 18% thick circular airfoil at zero incidence, Ma_∞=0.76 and rebased on chord of $11*10^6$ was assessed by Marvin et al. [3]. They observed that a thick shear layer forms downstream of the shock wave, which led to dramatic increase in turbulence kinetic energy and shear stress, both of which reached a maximum when the dissipation and diffusion of turbulence kinetic energy exceeded its production and then, decreased substantially.

Another investigation about the effect of turbulence intensity with low Reynolds numbers on NACA-0012 airfoil performance characteristics was undertaken by Li et al. [4], who studied the forces acting on the airfoil at Re=5300 for free stream turbulent intensities between 0.6% and 6%, varying the AOA from 0 to 30 degrees. Their study showed that even at low Reynolds numbers the effect of turbulence intensity is significant, showing an apparent increase of lift coefficient and decrease of drag coefficient at stall angle of 12 degrees. Their observation at stall angle for 0.6% and 6 % of free stream turbulence intensity showed that at low free stream turbulence intensities stall can be phased out; however at sufficiently large values of turbulence intensity stall conditions can be clearly identified at regular stall angle.

Huang et al. [5] studied the aerodynamic performance (lift and drag) of a cantilever NACA 0012 wing model at Re between $0.5\text{x}10^5$- $1.5\text{x}10^5$ with free stream turbulence intensity varying between

0.2% and 0.65%. They came to the conclusion that stall angle is delayed as turbulence intensity increases up to 0.45%. However, the effect of increasing free stream turbulence intensity on maximum lift coefficient became significant only at turbulence intensity above 0.45%.

Ahmed et al. [6] investigated the aerodynamic characteristics of wind turbine blades at Re between $3.8 \times 10^4 - 2 \times 10^5$ at AOA from -2 to 20 degrees. These authors found that both increase of AOA and free stream turbulence intensity lead to increase of turbulent kinetic energy production. The distribution of turbulence kinetic energy indicated the location of transition to turbulent region and how it shifted upstream as turbulence intensity was increased.

The present study aims to consider the effects of free stream turbulence intensity on the performance of the NACA 0012 airfoil; specifically, variation in drag, lift, turbulence kinetic energy, and energy dissipation characteristics will be studied.

3. Numerical model setup

The setup of the mathematical model is performed using the commercial-academic CFD software ANSYS-CFX. The main premises adopted here are: steady state, Newtonian fluid, incompressible and turbulent air flow. Therefore, the following Reynolds-averaged governing equations and boundary conditions were prescribed: (a) mass conservation; (b) Navier-Stokes (momentum) equations; (c) energy conservation; (d) constitutive relations between p, T and H; and (e) shear stress turbulence model. These equations are shown as follows:

$$\frac{\partial \rho}{\partial t} + \frac{\partial}{\partial x_i}(\rho u_i) = 0 \tag{3.1}$$

$$\frac{\partial \rho u_i}{\partial t} + \frac{\partial}{\partial x_j}(\rho u_i u_j) = -\frac{\partial p}{\partial x_i} + \frac{\partial}{\partial x_j}(\tau_{ij} + \tau_{ij}^R) + S_i \quad ; i = 1,2,3 \tag{3.2}$$

with:

$$\tau_{ij} = \mu\left(\frac{\partial u_i}{\partial x_j} + \frac{\partial u_j}{\partial x_i} - \frac{2}{3}\delta_{ij}\frac{\partial u_k}{\partial x_k}\right) \tag{3.3}$$

$$\tau_{ij}^R = \mu_t\left(\frac{\partial u_i}{\partial x_j} + \frac{\partial u_j}{\partial x_i} - \frac{2}{3}\delta_{ij}\frac{\partial u_k}{\partial x_k}\right) - \frac{2}{3}\rho k \delta_{ij} \tag{3.4}$$

Where Boussinesq´s assumption for the Reynolds-stress is adopted and δ_{ij} is the Kronecker delta function; μ is the dynamic viscosity, μ_t the turbulent eddy viscosity and k the specific turbulent kinetic energy.

The energy transport equation for the fluid in terms of the enthalpy is as follows:

$$\frac{\partial \rho H}{\partial t} + \frac{\partial \rho u_i H}{\partial x_i} = \frac{\partial}{\partial x_i}\left(u_j(\tau_{ij} + \tau_{ij}^R) + q_i\right) + \frac{\partial p}{\partial t} - \tau_{ij}^R\frac{\partial u_i}{\partial x_j} + \rho\varepsilon + S_i u_i + Q_H \tag{3.5}$$

where,

$$H = h + \frac{u^2}{2} \tag{3.6}$$

For air flow, the Ideal Gas state equation is prescribed to compute its density as a function of pressure and temperature:

$$\rho = \frac{p}{R_{air}T} \tag{3.7}$$

where,

$$R_{air} = \frac{R_{universal}}{Air\_Molecular\_Mass} \tag{3.8}$$

The Shear Stress Transport (SST) turbulence model consists of the well-known k-omega model near the wall and k-epsilon transport model in the free stream [7]:

$$\frac{\partial(\rho k)}{\partial t} + \frac{\partial(\rho u_j k)}{\partial x_j} = \rho P - \beta^* \rho \omega k + \frac{\partial}{\partial x_j}\left((\mu + \sigma_k \mu_t)\frac{\partial k}{\partial x_j}\right) \tag{3.9}$$

$$\frac{\partial(\rho\omega)}{\partial t} + \frac{\partial(\rho u_j \omega)}{\partial x_j} = \frac{\gamma}{\nu_t} P - \beta^* \rho \omega^2 + \frac{\partial}{\partial x_j}\left((\mu + \sigma_k \mu_t)\frac{\partial \omega}{\partial x_j}\right) + 2(1 - F_1)\frac{\rho \sigma_{\omega 2}}{\omega}\frac{\partial k}{\partial x_j}\frac{\partial \omega}{\partial x_j} \tag{3.10}$$

The constants ϕ of the new model are found using Eq. (3.13) below:

$$\phi = F_1 \phi_1 + (1 - F_1)\phi_2 \tag{3.11}$$

Eddy viscosity is calculated with

$$\nu_t = \frac{a_1 k}{\max(a_1 \omega; \Omega F_2)} \tag{3.12}$$

where Ω is the absolute value of the vorticity and also:

$$\begin{aligned} F_1 &= \tanh(\arg_1^4) \\ F_2 &= \tanh(\arg_2^2) \end{aligned} \tag{3.13}$$

$$\begin{aligned} \arg_1 &= \min\left[\max\left(\frac{\sqrt{k}}{0.09\omega y}; \frac{500 v}{y^2 \omega}\right), \frac{3.424 \rho k}{CD_{k\omega} y^2}\right] \\ \arg_2 &= \max\left(2\frac{\sqrt{k}}{0.09\omega y}; \frac{500 v}{y^2 \omega}\right) \\ CD_{k\omega} &= \max\left(1.712 \rho \frac{1}{\omega}\frac{\partial k}{\partial x_j}\frac{\partial \omega}{\partial x_j}, 10^{-20}\right) \end{aligned} \tag{3.14}$$

The respective boundary conditions are no-penetration/no-slip on the airfoil surface and remaining conditions are presented in table 1.

Table 1. Boundary conditions.

Parameter:	Value:
Re	9e6
Ma_∞	0.7
$T_{\text{air ideal gas}}$	288 [K]
$P_{\text{reference}}$	56867 [Pa]
AOA	1.49^0-16.49^0
Turbulence intensity	1%, 5%, 10%
Inlet	U= [U_x, U_y, 0]

Outlet	$P_{relative}$=0 [Pa]

The set of governing equations are discretized and solved using the finite volume (FV) method on a spatially structured computational mesh refined locally at the solid/fluid interfaces and in specified fluid regions where high gradients are expected.

The solution convergence criterion is reached when the residuals of normalized parameters are equal or smaller than $1.0\text{x}10^{-4}$. The mesh for the current project was provided by ANSYS Inc, which is depicted in Figure 2 and consists of 10,1520 nodes and 50,400 elements.

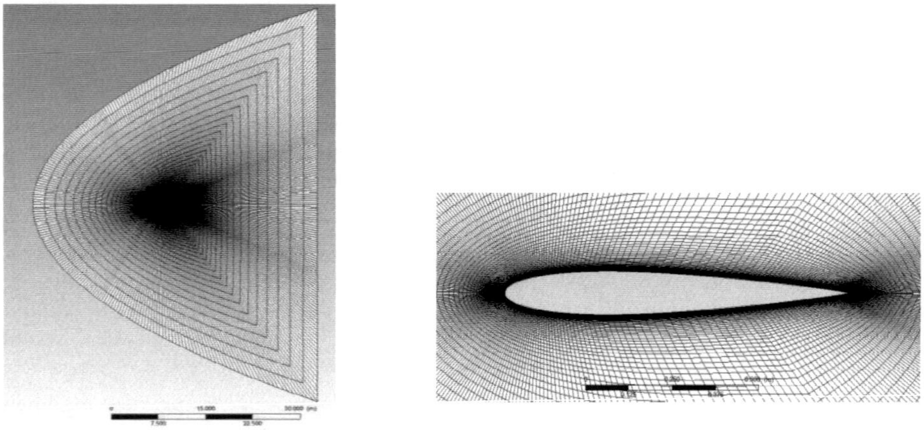

Figure 2. Computational mesh and its enlarged view.

The mesh was adopted from Ansys, therefore the only available option for mesh convergence study was to use the Mesh Adaption tool provided by the ANSYS software. This algorithm of mesh adaption, allows modifying certain regions within the mesh and consequently checking the corresponding solution for that adaption. Having adjusted the necessary criteria, simulations were run and results are summarized in Table 2.

Table 2. Mesh adaption setup parameters.

Test ID	Residual Target	Adaption criteria variable	Node factor	Maximum number of steps	Maximum iterations per step
1	1e-4	Density	2	5	50
2	1e-4	Velocity	2	25	250
3	1e-4	Density	2.5	25	300
4	1e-4	Density	1.5	25	300

21,000 iterations were performed until final mesh refinement was achieved. A final mesh of 108,642 nodes and 62,073 elements was obtained after subsequent steps of adaption, such that it satisfied the assigned residual target and number of iteration per step of adaption, as indicated in Table 2. Results of drag and lift were compared between initial and final meshes, finding a difference 0.2% for C_D and 0.3% for C_L coefficients. The difference was less than 1% and therefore, the adequacy of the original mesh structure was verified. Subsequently, all results are presented using the initial mesh.

A PC with an Intel® Xeon® E5620 CPU @ 2.40GHz with 4.00GB RAM running with two cores on a 64-bit Windows Operating System was used to perform the simulations. Each simulation with the chosen mesh took approximately 0.167 hours to fully converge.

SST turbulence model is validated using experimental results [8]. Results of Cp are presented in Figure 3, depicting a very good match between experiments and numerical results.

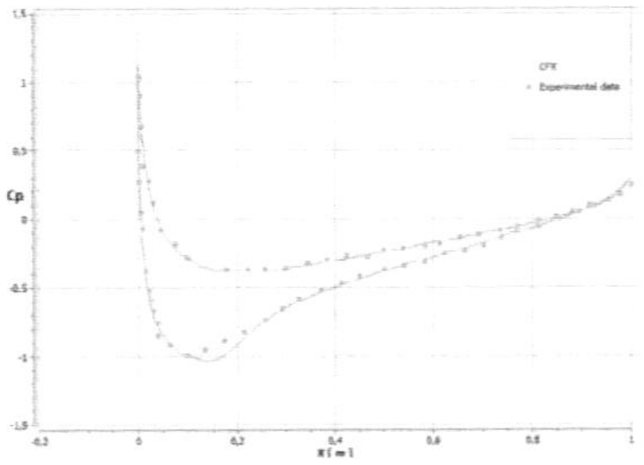

Figure 3. Comparison of pressure coefficient values between experimental data and ANSYS-CFX with SST turbulence model.

4. Numerical results

Values of C_D and C_L for different free-stream turbulence intensities were obtained. In order to observe the relation between the computed results, two separate figures were plotted. It was initially expected that there will be a reasonable deviation between the coefficients for different turbulence intensities, but the comparison revealed almost identical findings. According to figure 4, the drag coefficient C_D equally increases with the AOA growth. The same trend is observed for C_L parameters in figure 5, up to the point with AOA = 6.49 degrees, where stall occurs. The phenomenon of stall is recognized as it causes the boundary layer separation with subsequent dramatic fall of lift force and C_L. However, the later growth of AOA favors the lift coefficient recovery in all cases. In general, contrary to the expectations, the data closely fit into one common curve, no matter the intensity of free-stream turbulence being prescribed.

Results are partially understood by the fact that free-stream turbulence intensity, which is an independent variable for the simulations, does not affect the velocity mean value across the full domain. Obviously, the range of free stream turbulence levels does not have significant influence on the boundary layer and thus the lift and drag coefficients. However, the prescribed level of free-stream turbulence should have a direct impact on the magnitude of induced velocity fluctuations (i.e., turbulence kinetic energy) around the airfoil.

To see the real picture of the physical processes, it was decided to compare the turbulence kinetic energy and turbulence eddy dissipation fields among different cases. The 1% free-stream turbulence intensity case was taken as a benchmark, while others were compared with it. Simulations were performed for three different AOA: 1.49, 5.49 and 16.49 degrees. Turbulent kinetic energy plots at 1% and 5% turbulence intensity values are demonstrated on figure 6 and Figure 7. Numerical results revealed that there were significant differences in all situations, so the change in turbulence intensity had a finite effect on these turbulence fields. Nevertheless, changes are an order of magnitude smaller than absolute values of turbulence quantities and hence does not significantly affect the output parameter. For instance, the cases with low and medium turbulence intensities at 5.49 degrees AOA have order of magnitude 3, but their difference have and order of 2, as could be observed from Figure 8. This leads to the conclusion that despite the fact that some local turbulence quantities may be affected by changes of free-stream turbulence intensity, values of C_D and C_L are unaltered. Nevertheless, it is believed that the conditions just after the stall angle, require more research efforts due to expected much larger turbulence generation and therefore, it is proposed here for future work.

Finally, looking at the practical applications of the findings, the paper can assert that there is almost no effect of turbulence level variance within the range of 1 to 10% on the C_D and C_L parameters for NACA0012 airfoil.

IOP Conf. Series: Materials Science and Engineering **234** (2017) 012016 doi:10.1088/1757-899X/234/1/012016

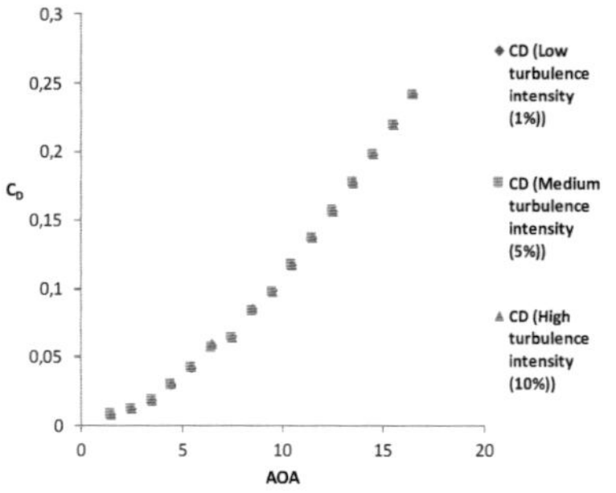

Figure 4. Drag coefficient values versus AOA.

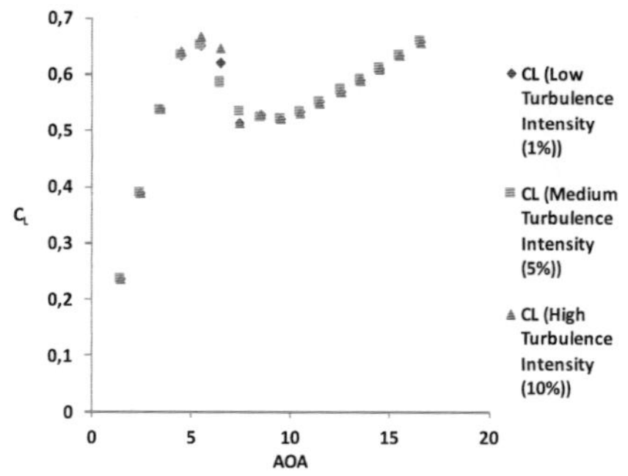

Figure 5. Lift coefficient values versus AOA.

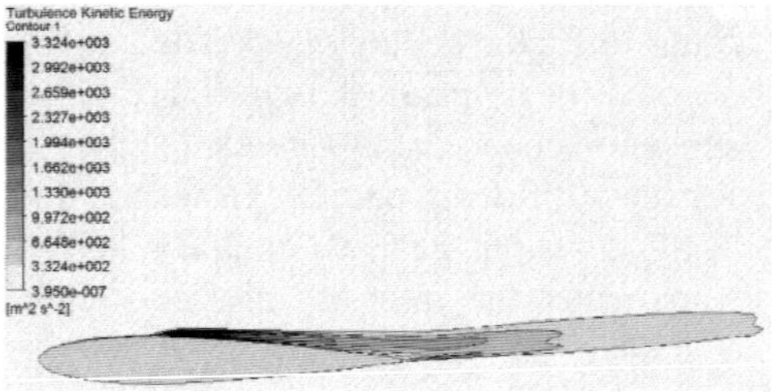

Figure 6. Turbulence KE for low turbulence intensity level (1%) at AOA 5.49 degrees

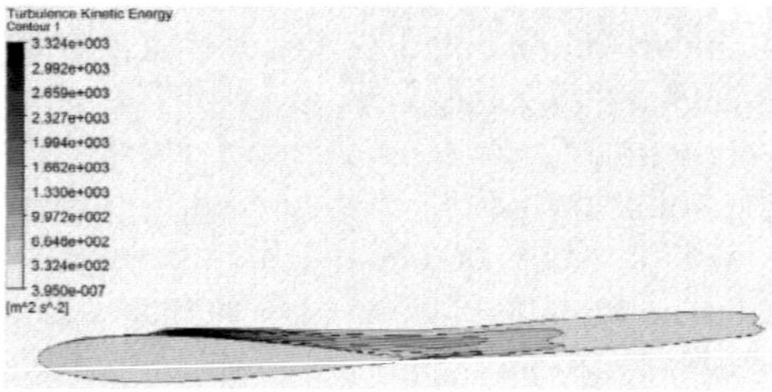

Figure 7. Turbulence KE for medium turbulence intensity level (5%) at AOA 5.49 degrees.

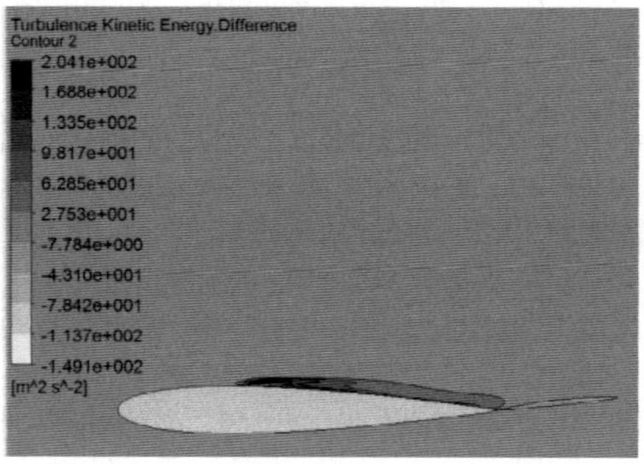

Figure 8. Difference between Turbulence KE for medium (5%) and low turbulence intensity level (1%) at AOA 5.49 degrees.

5. Conclusion

The current paper was devoted to the investigation of the effects of free-stream turbulence intensity on NACA-0012 airfoil in transonic conditions in the presence of a shock wave.

Changes in the turbulence kinetic energy and dissipation fields were observed, in a non-linear proportion with respect to free-stream turbulence intensity being prescribed (1-10%). However, maximum changes with respect to 1%-benchmark condition were always one order of magnitude smaller than turbulence field values.

Furthermore, almost negligible effects were observed on C_D and C_L coefficients under the same different levels of free-stream turbulence.

IOP Conf. Series: Materials Science and Engineering **234** (2017) 012016 doi:10.1088/1757-899X/234/1/012016

References

[1] Sahu, Kumar N and Imam S 2015 Analysis of transonic flow over an airfoil NACA0012 using CFD Intern. J. Innov. Sc. Eng. & Tech. 2 :4 ISSN 2348 – 7968.
[2] Yap T, Abdullah M Z and Husain Z 2002 The effects of turbulence intensity on the performance characteristics of NACA 0015 and Eagle 150 airfoil Proc. NSF Seminar.
[3] Marvin J G, Levy J L and Seegmiller H L 1980 Turbulence modeling for unsteady transonic flows AIAA J. 18:5.
[4] Li S, Wang S, Wang J and Mi J 2011 Effect of turbulence intensity on airfoil flow: numerical simulations and experimental measurements Appl. Math. Mech. 32: 8 pp. 1029–1038.
[5] Huang, Rong F and Lee H W 1999 Effects of free-stream turbulence on wing-surface flow and aerodynamic performance J. Aircraft 36:6.
[6] Ahmed, Rafiuddin M, Narayan S, AsidZullah M and Lee Y 2011 Experimental and numerical studies on a low reynolds number airfoil for wind turbine blades J. Fl. Sc. Tech. 6:3.
[7] Menter F 1994 Two-equation eddy-viscosity turbulence models for engineering applications AIAA J. 32(8) pp.1598-1605.
[8] Coakley T 1987 Numerical simulation of viscous transonic airfoil flows 25th AIAA Aerosp. Sc.

IOP Conf. Series: Materials Science and Engineering **234** (2017) 012017 doi:10.1088/1757-899X/234/1/012017

Investigating the Effects of Robot-Assisted Therapy among Children with Autism Spectrum Disorder using Bio-markers

Jaishankar Bharatharaj[1], Loulin Huang[1], Ahmed Al-Jumaily[1] , Mohan Rajesh Elara[2]and Chris Krägeloh[1]

[1]Institute of Biomedical Technologies, Auckland University of Technology, Auckland, New Zealand

[2]Engineering Product Development, Singapore University of Technology and Design, Singapore

E-mail: jaishankar.bharatharaj@aut.ac.nz

Abstract. Therapeutic pet robots designed to help humans with various medical conditions could play a vital role in physiological, psychological and social-interaction interventions for children with autism spectrum disorder (ASD). In this paper, we report our findings from a robot-assisted therapeutic study conducted over seven weeks to investigate the changes in stress levels of children with ASD. For this study, we used the parrot-inspired therapeutic robot, KiliRo, we developed and investigated urinary and salivary samples of participating children to report changes in stress levels before and after interacting with the robot. This is a pioneering human-robot interaction study to investigate the effects of robot-assisted therapy using salivary samples. The results show that the bio-inspired robot-assisted therapy can significantly help reduce the stress levels of children with ASD.

1. Introduction

The rapid increase in the number of children with autism around the world is becoming noticeable worldwide, putting a serious pressure on the human society to develop modern technological solutions to improve the lives of these individuals with autism. Centres for Disease Control and Prevention in the United States estimated that in 2014 about 1 percent of the world population (or about 74 million people) have some form of autism spectrum disorder [1]. Given that the medication to cure autism is yet to be found, many researchers are exploring new ways to help improve the lifestyle of these children. Robot therapy for improving the psychological, physiological, and social interaction abilities of children with autism is finding new directions in recent years. Particularly, biologically inspired robots were able to help increase the interaction interest and learning ability of children with autism. For instance, dog-like AIBO robot proved to be successful in establishing social interaction in children with autism [2]. In that study, the authors introduced a dog-like robot to eleven children and reported that the participants were much interactive with the robot than with a non-interactive dog-like toy. Paro, a seal-like robot has also reported success in improving social interaction abilities through various studies [3]-[5]. Probo is an imaginary animal-like robot used in autism therapy to improve social interaction and reduce stress [6]. Nao is a human-like intelligent robot initially used extensively in soccer playing robot competitions. In recent years, researchers have also used Nao robot in autism therapy giving various benefits such as improving learning, social interaction and joint exercise [7]-[8]. Animal-like robots have been developed by highlighting the benefits of their living counterparts in animal assisted therapies. The need for developing bio-inspired robots have been centred on the risk and challenges of using real animals. For instance, dogs were used extensively in animal-assisted therapy (AAT) with numerous reported benefits [9]-[10]. Likewise, there are many species that can provide therapeutic benefits to human being [11]-[13]. Nevertheless, hospitals and doctors show limited interest due to the fear of biting, allergy, and safety and ethical issues of using animals [14]-[16]. One such potential creature that can provide therapeutic benefits is parrot. After dogs, cats, and fish, parrots are thought to be the most common household pet [5]. Parrots are good in mimicking the human voice and other sounds. These birds are naturally able to speak and engage in two-way conversations with humans [17]. Pepperberg, an animal psychologist from the Harvard University has reported the intelligence of parrots through many studies. Particularly, her reports on Alex, the African grey parrot indicated that the parrots are not only able to speak but are able to build a two-way conversation. Pepperberg reported that Alex was able to identify 50 different objects and understand the concept of smaller, bigger, same, and different [18]. Parrots have been used in therapeutic settings and have provided remarkable benefits to the human society. For instance, parrots have been used in therapeutic setting to help people with mental or physical disorder. African grey parrot named Sadie was able to help his owner who had bipolar disorder with psychotic tendencies. Sadie helped his owner by repeatedly saying 'calm down' when needed [5]. In another setting, the parrot was used to assist patients with post-traumatic stress disorder among war veterans in the United States [19]. It is also reported that the parrots are helpful in therapeutic settings with psychiatric patients [20]. The distinct feature of talking ability in parrots is very relevant and helpful in improving learning and social interaction abilities of children with autism. Parrots have been used in many therapeutic settings including elderly care and autism therapy earlier. For instance, in Japan, an elderly care centre implemented parrot therapy and reported improvements in senses, such as sight, sound, and smell [21]. In the UK, an autistic child reported calming behaviour after interacting with a Caique parrot that was used as an emotional support animal [22]. Nonetheless, using parrots with children with autism is still a risky choice due to biting, allergy, and parrot-spread diseases. Specifically, parrot fever is spread by birds through the bacteria called, chlamydia psittaci [23].

To overcome these traditional challenges in parrot-assisted therapies, we have developed an artificial parrot robot, KiliRo that mimics its natural counterpart in therapeutic setting retaining the merits while overcoming the shortfalls. In this paper, we present evidence that the parrot-inspired robot that we developed can help reduce stress levels of children with ASD. To report reduction in stress levels, participants' urinary and salivary samples were analysed for protein and alpha amylase levels respectively. Investigating sociopsychological changes using non-invasive methods are gaining more attention in therapeutic settings during the recent decades. Especially, the studies investigating the effects of new therapies and methods are widely using biological samples to report changes in various aspects, such as psychological, physiological, and social interaction. Investigating substances such as protein and alpha-amylase can provide in-depth knowledge about the biological changes in participants during the experiments. Even though other sampling methods, such as blood test can provide the values of these substances, using non-invasive method is preferred to avoid pain and need for experts to collect samples. Urinary samples and salivary samples are found to be more appropriate and convenient methods to investigate protein and alpha amylase levels respectively. For such reasons, we used urinary and salivary samples of children to report changes in their stress levels. The results show that the participants' stress levels has decreased significantly after interacting with the robot.

2. Study setup

2.1. Participants

Open invitation was sent to all potential participants in identified schools through convenient sampling method three weeks before the study. Parents were given two weeks to respond to the invitation. A total of 19 invitations were sent to which 14 parents' responded that they would be willing for their children to be involved in this experiments. A meeting was conducted with interested parents 7 days prior to the experiments and a request was made to complete the Childhood Asperger Syndrome Test (CAST) questionnaire to identify the autism behaviour in their children. CAST involves a set of thirty-nine 'yes' or 'no' questions captured in a questionnaire to evaluate autism behaviour in children [5]. Children scoring 15 and above were classified to have symptoms of autism. Out of 17 children five were excluded from the study due to low score in CAST questionnaire and another two were excluded due to severe autism behaviour as judged by the paediatrician and the child psychologist. Finally, our experiments were conducted with seven participants over a duration of 49 days. Participating children were between the age group of 6 to 16 years and had no exposure to robots. The mean age of participating children was 10.0 years with a standard deviation of 3.51. To protect the identity of participants, their urinary and salivary samples were labelled as Subject1 through 7.

2.2. Study setup
Participating children were exposed to the robot for a minimum duration of 60 minutes each day over 49 days. Participants were allowed to interact and carry the robot freely. This study was performed with no compulsory attendance requirements for the participants. Nevertheless, the researcher and volunteers were present every day throughout the study adhering to the maximum working hours as instructed by employment rights in New Zealand and India. Four sessions were conducted each day with a minimum of 15 minutes per session. During the first session, the robot was placed on a table with the dimension of 3x3 square feet. The participants were allowed to have open interaction with the KiliRo robot. The second and third sessions involved activities around learning abilities of the robot. Letters of the alphabet and numbers were shown to the robot, and its ability to recognize them was demonstrated in front of participating children. During the fourth session, the robot was operated to move forward and backward while talking to the participants about its capabilities.

2.3. Urinary and salivary test
Literature points to non-invasive methods for assessing human psychological and social behavioural processes as a highly convenient and effective approach in many clinical settings. Particularly, these methods are found to be very useful in research studies involving special needs children. For instance, Wada and Shibata investigated the 17-KS-S and 17-OHCS hormone levels in elderly using urinary samples to evaluate the stress level in elderly while interacting with a seal-like robot [24]. In this study 12 participants aged between 67 and 89 years. The participants were assessed by mini mental state examination (MMSE) to investigate their mental cognitive state. During the study, the participants were allowed to interact with the PARO robot freely and the reactions of the participants' vital organ were investigated using urinary samples. The study reported in the 17-KS-S/17-OHCS ratio after two weeks of interaction with the robot.

However, there are other hormones in the human body whose proportional relationships to psychological changes are better quantified. It is identified that higher protein levels in urinary sample of higher stress levels in subjects [25]. Previous studies have reported that a person's alpha-amylase level generally increases during stressful situations [26]. Sympathoadrenal medullary (SAM) activity is examined in alpha-amylase test and increased alpha-amylase levels serves as an indicator of increased stress levels [27].

For our experiments presented in this paper, we used urinary and salivary tests to evaluate the biological changes in children with autism before and after interacting with our parrot-inspired robot, KiliRo. This is a pioneering human robot interaction studies to use salivary test for validating the effects of robotic engagement with autistic children. Urinary sample was used to measure protein levels in participating children and salivary sample was used to measure the alpha-amylase levels. Urinary and salivary samples of participating children were collected by their parents and was handed over to the laboratory technician on the first day of the study, before introducing our KiliRo robot. The first samples were collected on the first day of the study before the participants have interacted with the robot, while the second samples were collected during the last week of the study to evaluate the changes in stress levels after interacting with KiliRo robot for over 7 weeks. Both the urinary and salivary samples were at the study venue between 9 a.m. and 11 a.m.

3. KiliRo

The schematic diagram of our KiliRo robot is illustrated in Fig. 1. The body frame of our robot was completely fabricated using a 3D printer. PLA (Polylactic Acid) and TPU (Thermoplastic Polyurethane) materials were used in printing the robot. The robot consists of one head with two eyes, a body with two wings and one tail, and two legs. The head has one degrees of freedom enabling it to turn left and right. The beak and each wing has one degree of freedom to move up and down. The legs of the robot have two degrees of freedom allowing to move forward/backward and turn left/right. Two Ai-ball wireless cameras and one USB camera were used to achieve almost 270-degree field of vision. In addition to achieve real-time video transmission, the robot platform is capable of recognizing alphabets and number through use of QR codes. The robot is capable of recognizing English alphabets from A through Z and Arabic numerals from 0 through 9 using the QR code scanning algorithm. Seven touch sensors were used to recognize the tactile interaction preferences of children with KiliRo robot. These sensors were connected in serial to ensure touch of any sensor would enable the LED light on the robot to glow encouraging the children for more physical interaction with KiliRo. When a number or an alphabet was shown to the robot, it reads the QR code pasted in the model and the corresponding text is converted into speech using text-to-speech module.

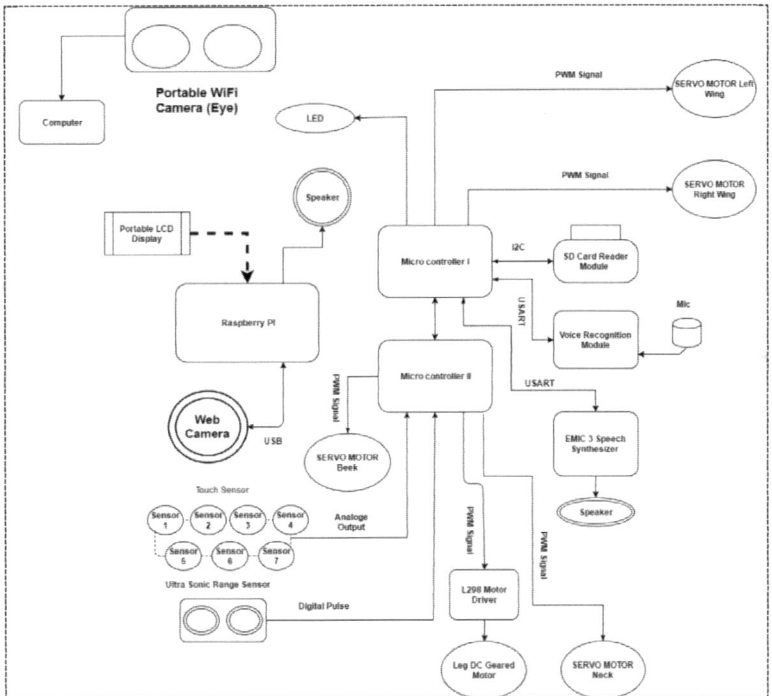

Figure 1. KiliRo – Schematic diagram.

Two Arduino family of microcontrollers and one raspberry pi-3 boards were used to synthesize a range of autonomous behaviours on KiliRo including, speech recognition, object recognition, real-time video transmission, text-to-speech, speaker, microphone, and locomotion gaits. Speech recognition module enabled the robot to recognize ‘yes’ and ‘no’ from the participants’ and act accordingly. This feature encouraged the participating children to have verbal communication with our KiliRo robot. The speech recognition feature was used to initiate and terminate the robot’s speaking module. For instance, when ‘Start’ was recognized, the robot started introducing itself and when ‘Stop’ was recognized, the robot stopped talking. Bipedal mechanism was used for synthesizing the robot locomotion. The robot was able to walk forward, backward, and turn left and right using a pair of DC motors powered using a 2000mAh lithium polymer battery. For the experiments presented in this study, KiliRo's locomotion was achieved using an android application installed on a smart phone. Fig. 2 illustrates our KiliRo robot used in this study.

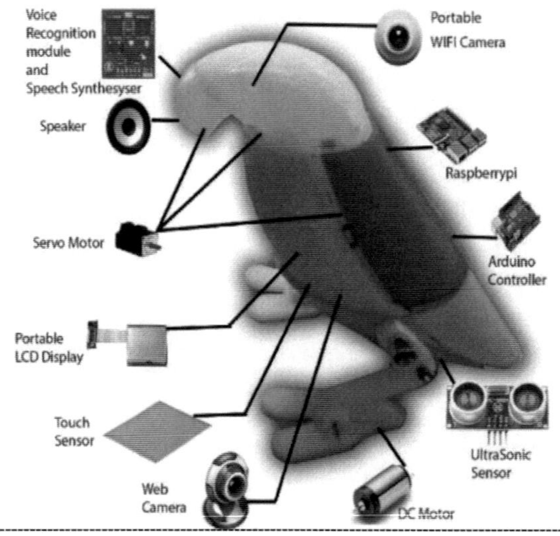

Figure 2. KiliRo robot.

4. Results

Table.1 show the protein and alpha-amylase levels of participants before and after interacting with the robot.

Table 1. Urinary and salivary test results.

Participant number	Protein (mg/dL) (Reference range 0 – 20)		Alpha-amylase (U/L) (Reference range 59 – 401)	
	Before	After	Before	After
1	69.7	19.8	564	98
2	22.5	8.4	179	76
3	29.5	18.8	348	132
4	32.9	25.0	218	121
5	19.0	10.4	113	131
6	22.5	13.4	177	81
7	85.2	26.0	541	420

In our trials, we observed that the average protein level of participating children dropped from 40.19 to 17.40 after interacting with our KiliRo robot. These changes were statistically significant according to a related-samples Wilcoxon Signed Ranks Test ($z=-2.37$, $p<.05$). The obtained results clearly demonstrate the efficacy and validity of our developed KiliRo robot in lowering the stress level of children with autism through close human robot interactions.

It is also noted that the alpha-amylase levels have reduced in the second sample test report, from a mean of 305.71 to 151.29. Despite the fact that the level for participant 5 increased at the second time point, the overall reduction in the alpha-amylase levels of the group was statistically significant (related-samples Wilcoxon Signed Ranks Test, $z=-2.20$, $p<.05$).

The results points to high degree of reduction in stress levels of children with ASD through reduced protein and alpha-amylase levels. All children participated in the study has shown reduction in both substances investigated. Such results are more encouraging to explore robot-assisted therapy among children with ASD.

5. Conclusions

We have developed a parrot-inspired robot, KiliRo and validated the merits of our robot in reducing stress levels and improving physiological conditions with a group of seven children identified with autism over a period of 49 days. We investigated the changes in stress levels of participating children after interacting with the robot through urinary and salivary tests. This is the first time in human-robot interaction study to use salivary sample investigating the changes in stress levels. The urinary and salivary test reports from our study indicated reduction in stress levels among the participants after interacting with our KiliRo robot.

This study has several limitation. Although the results show significant reduction in stress levels of participating children, the study did not have a control group to compare the results. Even though protein and alpha-amylase levels indicate reduction in stress levels of participants after interacting with the robot, several other factors that may have influenced these chemical substances such as, medication, body temperature, personal or cultural activity, and sample taken time. Also the sample size was too low to generalize the findings of the study. Hence, the results of the study should be viewed carefully, and replication with large sample size and control group is needed. With this, we foster to run an extended study with a large target population of minimum 100 people with a control group to compare the findings. Another possibility of future work is to conduct a study with longer duration of over several months. The third area of our interest is to build on our present work and run cross country study at three different locations; India, Singapore, and New Zealand to compare the results. Finally, we intend to extent the autonomous and locomotion capabilities of our KiliRo robot with increased autonomous functions.

References

[1] Autism Rescue Angels [Internet]. Autism Facts and Statistics. Autism Rescue Angels; 2015 [cited 2016Oct24]. Available from: http://autismrescueangels.org/resources/

[2] Stanton CM, Kahn PH, Severson RL, Ruckert JH, Gill BT. Robotic animals might aid in the social development of children with autism. In *Human-Robot Interaction (HRI),* 2008 3rd ACM/IEEE International Conference on 2008 Mar 12 (pp. 271-278). IEEE.

[3] Yu R, Hui E, Lee J, Poon D, Ng A, Sit K, Ip K, Yeung F, Wong M, Shibata T, Woo J. Use of a therapeutic, socially assistive pet robot (PARO) in improving mood and stimulating social interaction and communication for people with dementia: Study protocol for a randomized controlled trial. *JMIR research protocols*. 2015 Apr; 4(2).

[4] Wada K, Shibata T. Living with seal robots—its sociopsychological and physiological influences on the elderly at a care house. *IEEE Transactions on Robotics.* 2007 Oct; 23(5):972-80.

[5] Bharatharaj J, Huang L, Al-Jumaily A. Bio-inspired therapeutic pet robots: Review and future direction. In Information, Communications and Signal Processing (ICICS), 2015 10th International Conference on 2015 Dec 2 (pp. 1-5). IEEE.

[6] Vanderborght B, Simut R, Saldien J, Pop C, Rusu AS, Pintea S, Lefeber D, David DO. Using the social robot probo as a social story telling agent for children with ASD. Interaction Studies. 2012 Jan 1; 13(3):348-72.

[7] Shamsuddin S, Yussof H, Ismail L, Hanapiah FA, Mohamed S, Piah HA, Zahari NI. Initial response of autistic children in human-robot interaction therapy with humanoid robot NAO.

InSignal Processing and its Applications (CSPA), 2012 IEEE 8th International Colloquium on 2012 Mar 23 (pp. 188-193). IEEE.
[8] Tapus A, Peca A, Aly A, Pop C, Jisa L, Pintea S, Rusu AS, David DO. Children with autism social engagement in interaction with Nao, an imitative robot: A series of single case experiments. Interaction studies. 2012 Jan 1; 13(3):315-47.
[9] Richeson NE. Effects of animal-assisted therapy on agitated behaviors and social interactions of older adults with dementia. American Journal of Alzheimer's Disease & Other Dementias®. 2003 Nov; 18(6):353-8.
[10] Perkins J, Bartlett H, Travers C, Rand J. Dog - assisted therapy for older people with dementia: A review. Australasian Journal on Ageing. 2008 Dec 1; 27(4):177-82.
[11] Knisely JS, Barker SB, Barker RT. Research on benefits of canine-assisted therapy for adults in nonmilitary settings. US Army Med Dep J. 2012 Apr 1:30-7.
[12] Fine AH, editor. Handbook on animal-assisted therapy: Theoretical foundations and guidelines for practice. Academic Press; 2010 Sep 21.
[13] Levinson BM. Human/companion animal therapy. Journal of Contemporary Psychotherapy. 1984 Sep 1; 14(2):131-44.
[14] Iannuzzi D, Rowan AN. Ethical issues in animal-assisted therapy programs. Anthrozooes. 1991 Sep 1; 4(3):154-63.
[15] Bharatharaj J, Kumar SS. Considerations in Autism therapy using robotics. In Computing, Communications and Networking Technologies (ICCCNT), 2013 Fourth International Conference on 2013 Jul 4 (pp. 1-5). IEEE.
[16] Wada K, Shibata T. Robot therapy in a care house-results of case studies. In Robot and Human Interactive Communication, 2006. ROMAN 2006. The 15th IEEE International Symposium on 2006 Sep 6 (pp. 581-586). IEEE.
[17] Bharatharaj J, Huang L, Mohan RE, Al-Jumaily A, Krägeloh C. Robot-Assisted Therapy for Learning and Social Interaction of Children with Autism Spectrum Disorder. Robotics. 2017 Mar 14; 6(1):4.
[18] Smith D. A thinking bird or just another birdbrain. The New York Times. 1999 Oct: 1.
[19] Parrots Used in PTSD Therapy for War Veterans. Retrieved September 10, 2016, from http://www.npr.org/templates/story/story.php?storyId=11989027.
[20] Haw C. Parrots as therapy for psychiatric patients. The Psychiatrist. 2007 Apr 1; 31(4):154-5.
[21] Therapy Birds: "Emotional Support Animal. Retrieved September 10, 2016, from http://www.birdchannel.com/therapy-pet-birds.aspx
[22] Parrots As Animal Assisted Therapy [Internet]. For Parrots: Posters for Parrot Advocates. 2013 [cited 2017May10]. Available from: http://www.forparrots.com/2013/09/07/parrots-as-animal-assisted-therapy/.
[23] Smith KA, Bradley KK, Stobierski MG, Tengelsen LA. Compendium of measures to control Chlamydophila psittaci (formerly Chlamydia psittaci) infection among humans (psittacosis) and pet birds, 2005. Journal of the American Veterinary Medical Association. 2005 Feb 15; 226(4):532-9.
[24] Wada K, Shibata T. Living with seal robots in a care house-evaluations of social and physiological influences. In Intelligent Robots and Systems, 2006 IEEE/RSJ International Conference on 2006 Oct 9 (pp. 4940-4945). IEEE.
[25] Phillips N. 24-hour Urine Protein Test [Internet]. Healthline. Healthline Media; 2016 [cited 2017May10]. Available from: http://www.healthline.com/health/24-hour-urine-protein
[26] van Stegeren A, Rohleder N, Everaerd W, Wolf OT. Salivary alpha amylase as marker for adrenergic activity during stress: effect of betablockade. Psychoneuroendocrinology. 2006 Jan 31; 31(1):137-41.
[27] Rohleder N, Nater UM, Wolf JM, Ehlert U, Kirschbaum C. Psychosocial stress - induced activation of salivary alpha - amylase: an indicator of sympathetic activity?. Annals of the New York Academy of Sciences. 2004 Dec 1; 1032(1):258-63.

IOP Conf. Series: Materials Science and Engineering **234** (2017) 012018 doi:10.1088/1757-899X/234/1/012018

Wing Rib Stress Analysis and Design Optimization Using Constrained Natural Element Method

Mohamed Amine Bennaceur[1,a], **Yuan-ming Xu**[1] **and Hemza Layachi**[1]

[1]School of Aeronautic Science and Engineering, Beihang University, Beijing 100191, China

[a] amine_m@qq.com

Abstract. This paper demonstrates the applicability of a novel meshless method in solving problems related to aeronautical engineering, the constraint-natural element method is used to optimize a wing rib where it present several shape of cut-outs deals with the results findings we select the optimum design, we focus on the description and analysis of the constraint-natural element method and its application for simulating mechanical problems, the constraint natural element method is the alternative method for the finite element method where the shape functions is constructed on an extension of Voronoi diagram dual of Delaunay tessellation for non-convex domains.

1. Introduction

The wing ribs are structural elements of the airplane; with the longeron they form the skeleton for the wing skin where their functions are to maintain its shape. The ribs do not participate in the strength of the aircraft as a whole, but transmit the forces encountered by the skin to the longeron. They are subjected to loading systems. External loads applied in the plane of the rib produce a change in shear force in the wing across the rib [1]. The geometry may also include holes to reduce the weight of the wing without altering its strength. The ribs are cut to allow to fuels or equipment to pass through it [2], [3]. In this paper the wing rib of a light aircraft are analyzed with different cut-outs design.

The design of a new wing rib requires stress analysis. In engineering, stress analysis is a tool rather than a goal; the aim is to determine the stress and to predict the failure in materials subjected to forces or increase the strength of the structure without increasing the weight. Stress analysis can be performed using conventional and analytical mathematical techniques, an experimental testing or computational simulation.

Widely used in computational mechanics the finite element method is considered as the robust method to analysis mechanical phenomena [4] but this method failed to analyze some case such as large transformations problems, a localization of initiation and propagation of a crack, displacement of interfaces between solid and fluid. In the last few years there has been an increasing interest in a novel numerical method alternative to the finite element method, meshless methods have been developed with the aim to solve problems associated mesh steps linking to the finite element method. meshless method construct the interpolating around a set of point that means no connection between nodes and no contract of shape of the element, the refining of mesh or in this case a set of point is automatically by admitting new points; however those methods suffer from a major problem where the imposition of boundary conditions is very complex and the support of the shape function must be contained a sufficient number of neighboring nodes.

A new meshless method is developed to solve this problem by using Galerkin type discretization and diagram of Voronoi. the Natural element method merge the advantage of the classic method (finite element method) and the meshless method using direct application of the boundary conditions of the Dirichlet type. It is enough to impose the values corresponding to the nodes of the edge in aim to obtain the desired solution on the edge whole [5].

On the other hand, on a non-convex edge, this property is no longer verified. In the case of points on either side of a fort non-convex domain may be natural neighbors, and therefore have an undesirable mutual influence. To solve the problem of non-convex edges, the simplest solution is to use the constrained Delaunay triangulation associated with a visibility criterion. The visibility criterion makes it possible to avoid points situated on either side of a non-convex domain each other [6], [7]. The NEM (Natural Element Method) then becomes C-NEM (Constraint- Natural Element Method).

2. Construction of function shape natural element

Introduced by Sibson, the notion of natural neighbours and natural neighbours interpolation is based on a geometric constructions known as the Voronoi diagram. The Voronoï diagram is the topological dual of the Delaunay tessellation [8]. Let's have a cloud of separated irregular nodes $N = \{n_1, n_2, ..., n_N\}$ on the region R^2. The first-ordre Voronoi diagram divides the plane into a set of cell T_i, one for each node n_i [6], [9] see figure 1 (a). Then the Voronoi diagram is defined by:

$$T_i=\{x \in R^2 : d_i<d_j \ \forall j \neq i\} \quad (1)$$

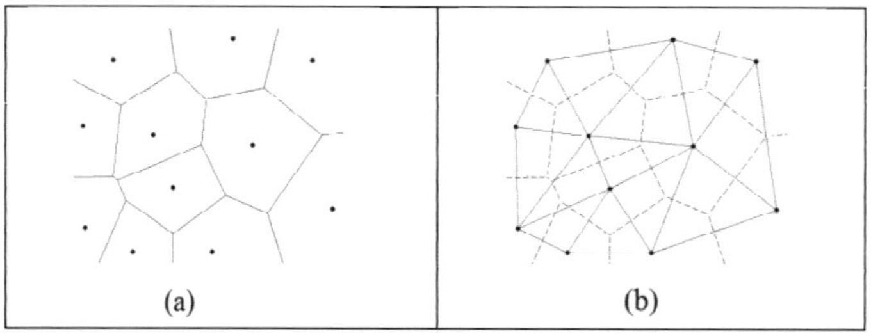

Figure 1. (a) Voronoi diagram (b) Delaunay triangulation

Where x is the coordinate of a node n and d_i, d_j are the distance between n and n_i, n_j respectively; next the Delaunay tessellations are obtained by linking the nodes whose Voronoi cells have a common boundaries see figure 1 (b).

On the way to calculate the value of the Sibson interpolation, let's added a point x to pervious region see figure 2, by extension it is possible to define a higher order Voronoi diagram (To the order k, k> 1) [10]. The second-order Voronoi cells T_{ij} is defined as:

$$T_{ij}=\{x \in R^2 : d_i<d_j<d_k \ \forall k \neq i,j\} \quad (2)$$

Where ni and nj are the first and the second closest nodes to the point x.

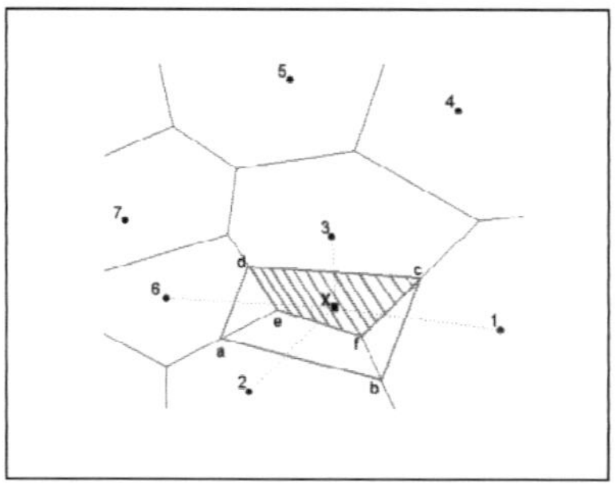

Figure 2. First and second order Voronoi cells about x

Considered κ a Lebesgue measure (area 2D) the Sibson interpolation is given by:

$$\Phi_i(x) = \frac{\kappa_{xi}}{\kappa_x} \quad (3)$$

$$\text{And} \quad \kappa_x = \sum_{i=1}^{n} \kappa_{xi} \quad (4)$$

Where κ_{xi} is the cell of the second voronoi diagram of a points *x*.

$$\Phi_i(x) = \frac{Area\ (cdef)}{Area\ (abcd)} \quad (5)$$

Braun and Sambridge use this interpolation in a Galerkin type discretization to solve partial differential equation and give it the name of Natural element method [5].

Consider a vector-valued function: $u(x) \subset R^2 \rightarrow R^2$ the interpolation scheme is in the form:

$$u(x) = \sum_{i=1}^{n} \Phi_i(x) u_i \quad (6)$$

Where u_i ($i = 1, 2, \dots, n$) are the vectors of nodal displacements at the *n* natural neighbours, and $\Phi_i(x)$ are the shape functions associated with each node.

3. Extension of natural element method for non-convex domain

In the case of a fort non-convex domain several points in the either sides may be neighbours and therefore have an undesirable mutual influence see figure 3 (a). Yvonnet and al [6] via a constrained Voronoi diagram a dual of constrained Delaunay triangulation results from the application of a visibility criterion define the interpolation functions for a non-convex domain and keeping the main properties of the NEM interpolation. A point *x* is visible from a node n_i inside the domain if a segment drawn between n_i and *x* does not cross the boundary of the domain Γ see figure 3 (b).

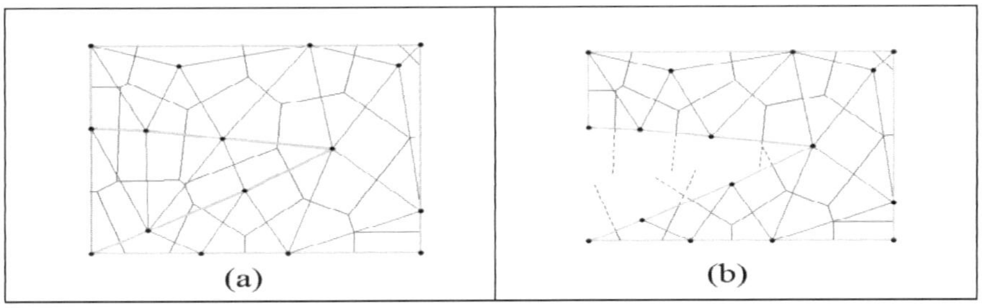

Figure 3. (a) Voronoi diagram and Delaunay triangulation for a non-convex domain.
(b) constrained-Voronoi diagram and Delaunay triangulation for a non-convex domain.

If the visibility criterion is introduced in the NEM, the natural neighbours become constrained natural neighbors. Thus, the new functional approximation can be written as:

$$u(x) = \sum_{i=1}^{v} \phi_i^C(x) u_i \quad (7)$$

Where v is the number of natural neighbours visible from point *x* and ϕ_i^C is the constrained natural neighbor shape function [6].

4. Case study

For this task, we use the constrained natural element method to optimize a wing rib; this wing rib is attached to the front spar, rear spars and wing skins see figure 4 (a). In this particular case the weight of the rib is imposed, we want to increase the strength of rib without increasing its weight.

IOP Conf. Series: Materials Science and Engineering **234** (2017) 012018 doi:10.1088/1757-899X/234/1/012018

The rib is made from Aluminium 7075-T73 a very high strength material used for highly stressed structural parts, the material properties are shown in the table 1.

Table 1. Material properties.

Properties	Values	Properties	Values
Density	2810 kg m^{-3}	Shear Modulus	26.9 GPa
Modulus of Elasticity	72 GPa	Tensile Strength, Yield	435 MPa
Poisson's Ratio	0.33	Shear Strength	300 MPa

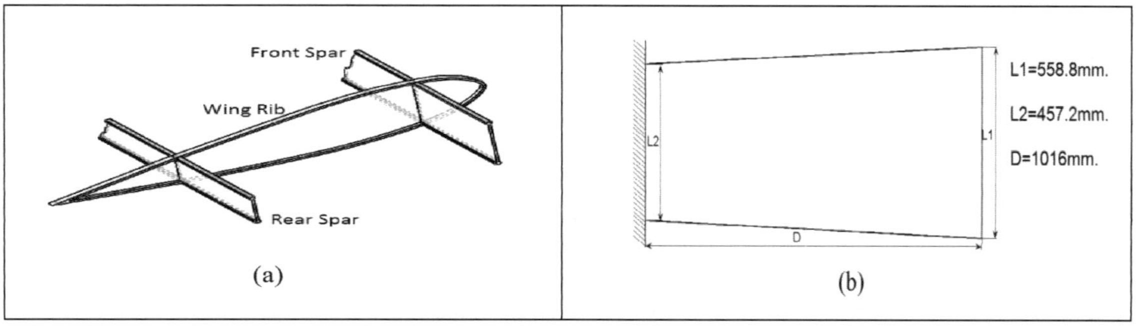

Figure 4. Wing ribs synoptic scheme. (a) Wing box. (b) Wing rib design variables

To determine the optimal configuration, The geometry will be simplified by ignoring the curvature of the rib edges and we assume that the front spar loads the rib in shear and the wing rib is built in the rear spar, the shear forces q=10 KN see figure 4 (b). We will analyze three different designs where the shapes of the holes are modeled in several form as shown in the figure 5.

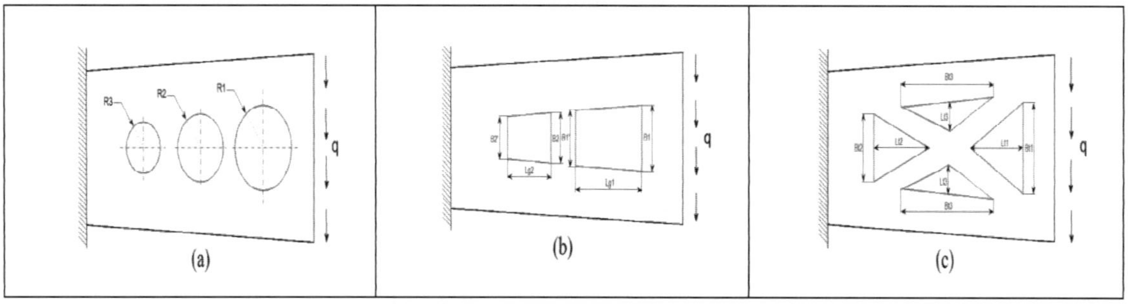

Figure 5. Modeling of rib structure. (a) With circular holes. (b) With isosceles trapezoid holes. (c) With triangular holes

The first design include three circular holes with a varied diameter where R1=127mm, R2=101.6mm and R3= 76.2mm see figure 4 (a). The second design consist two isosceles trapezoid holes where the first isosceles trapezoid had for bases B1=215mm, B1'=183.6mm and the legs Lg1=286.6mm, The second isosceles trapezoid had for bases B2=165mm, B2'=140.9mm and the legs Lg2=192.4mm see figure 4 (b).The last design represent four triangular holes with varying lengths and bases where two are isosceles triangle and have for bases Bt1=274mm, Bt2=203.2 mm and lengths Lt1=236.7mm, Lt2=249.4mm, the last two triangle had the same shape where the base Bt3=413mm and the lengths Lt3=88.5mm see figure 4 (c).

5. Governing equation

Let's consider a linear elastic body, homogeneous and isotropic, which undergoes small deformations and small displacements. It occupies a domain *V*. in which we denote the surface boundary by *S* and subjected to the body forces b_i in *V*, to the surface traction t_i on S_t and displacement u_i imposed on S_u. Where $S_u \cup S_t = S$ (see figure 6).

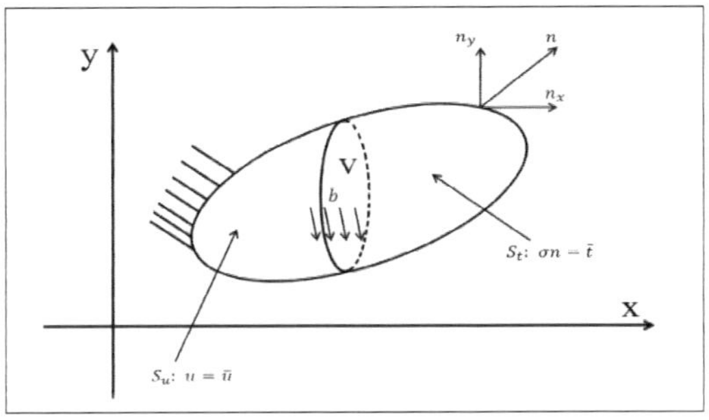

Figure 6. Elastic body undergoes several forces

The equation of equilibrium described by the linear elastic body is given in indicial form as follow:

$$\sigma_{ij,j} + b_i = 0 \tag{8}$$

The constitutive relation is given according to the stress – strain relation (Hooke's law) in matrix form as:

$$\{\sigma\} = [C]\{\varepsilon\} \tag{9}$$

And [C] is the material property matrix.

The principle of virtual displacements (PVD) presents an indirect way to express the equilibrium of a body. This PVD states that the total internal virtual work is equal to the total external virtual work.

$$\int_v \{\sigma\}^T \{\bar{\varepsilon}\} dv = \int_v \{P_v\}^T \{\bar{u}\} dv + \int_{S_T} \{P_s\}^T \{\bar{u}\} ds \tag{10}$$

According to this, substituting the approximation functions by the C-NEM scheme (7) in the preceding equation (10), we obtain a system of linear equation, the governing equations is given in this form:

$$[K]\{\delta\} = \{F\} \tag{11}$$

$\{\delta\}$ is the vector containing the nodal displacements, with the stiffness matrix [K] is given by:

$$[K] = \int_v [B]^T [C][B] dv \tag{12}$$

And
$$F = \int_v [N]^T \{b\} dv + \int_{S_T} [N]^T \{\bar{t}\} ds \tag{13}$$

[N] is the matrix containing the shape functions

$$[N] = \begin{Bmatrix} \phi_1 & 0 & \phi_2 & 0 & \cdots & \phi_N & 0 \\ 0 & \phi_1 & 0 & \phi_2 & \cdots & 0 & \phi_N \end{Bmatrix} \tag{14}$$

And [B] is the matrix containing the derivatives of shape functions C-NEM.

$$[B] = \begin{Bmatrix} \phi_{1,x} & 0 & \phi_{2,x} & 0 & \cdots & \phi_{N,x} & 0 \\ 0 & \phi_{1,x} & 0 & \phi_{2,x} & \cdots & 0 & \phi_{N,x} \\ \phi_{1,y} & \phi_{1,x} & \phi_{2,y} & \phi_{1,x} & \cdots & \phi_{N,x} & \phi_{N,x} \end{Bmatrix} \tag{15}$$

6. Numerical simulation and discussion

The objective is to determine the stress field in the structure. Concluding the maximum stress and maximum vertical displacement of the rib where the maximum stress must be below the yield stress of the rib material and the maximum vertical movement of the rib should not exceed 2.5mm (design criterias). This is a selective study of the best performing configuration.

IOP Conf. Series: Materials Science and Engineering **234** (2017) 012018 doi:10.1088/1757-899X/234/1/012018

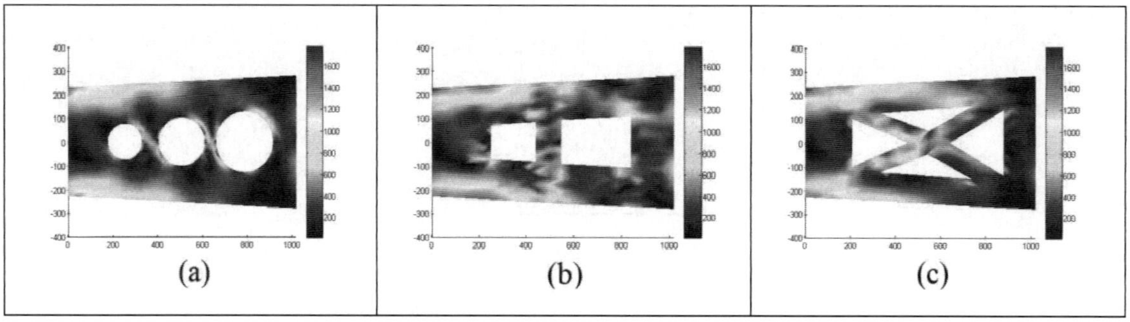

Figure7. Von-mises stress distributionin the wing rib (a) with three circular holes (b) with two trapezoid holes (c) with four triangular holes.

Figure 7 represent the simulation results of the Von-mises stress distribution for the three proposed design, from this figure we can notice that the stressed area are located in the upper rear spar. By applying a shear load, for the given design varieties, the Von-mises stress around the cut-outs does not represent a significant intensity where changing the shape of the holes, the stressed zone still located on the rear spar, the form of the cut-out does not influence the distribution of stress whether it is circular, triangular or trapezoid.

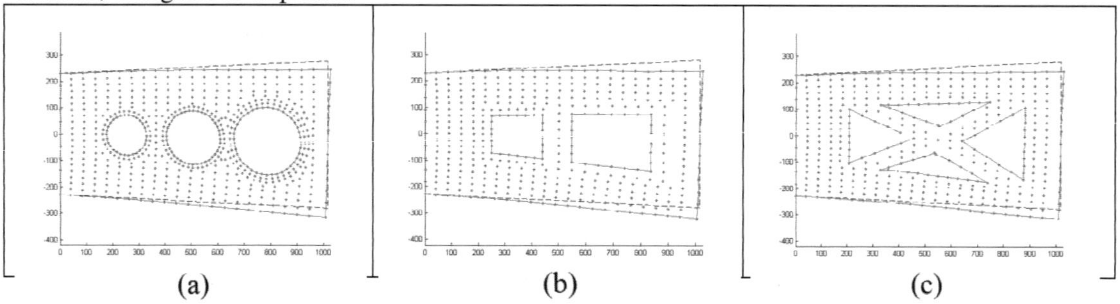

Figure 8. Displacement field of the wing rib structure
(a) with three circular holes (b) with two trapezoid holes (c) with four triangular holes.

Figure 8 shows the displacement field of the wing rib structure under shear forces, we realise no considerable difference between the three designs except for the trapezoid form. It is explained for the reason that the frontal opening is large and endures such displacement. The overall simulation results are shown in the table 2.

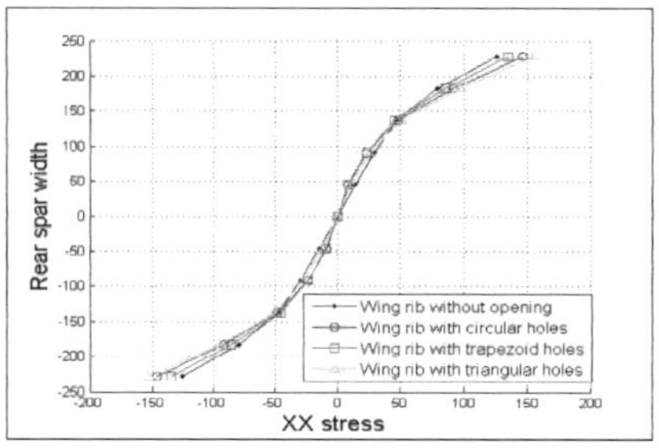

Figure 9. XX stress distribution along the rear spar width

Table 2. Stress and displacement values.

	Von-Mises stress (Mpa)	Max stress (Mpa)	Max displacement(mm)
Configuration A	1.6113	1.4654	2.1996
Configuration B	1.3812	1.3474	2.3888

Configuration C	1.7841	1.5515	2.1935

As concluded from the previous section, the form of the holes does not affect the destabilization process around the circumference of the cut-out itself though; the initiation of cracks is located in the upper rear spar area. The form of the hole take part to sustain the load and diminish the initiation of cracks as it is the case of trapezoid form, the stability performance are given at maximum comparing to a rib without opening see figure 9.

7. Conclusion

In this paper we used the constrained natural element method to optimize the cut-out in the wing ribs of a light-aircraft by adopting three different configurations. This novel work shows that changing the form of the cut-out will increase the strength of the wing rib. For this particular case, the configuration with trapezoid holes leads to optimal stability performance to sustain shear load additionally, it gives a big opening, which allow passing and maintaining the wing interior equipments. The constrained natural element method was applied successfully to solve problem in analyses of stress and design of new structure in aeronautic engineering and can be the alternative of the finite element method in commercial software where, the proposed approach reveals various applications in mechanical engineering. As a perspective to this work, extend the study to define the influence of buckling on wings ribs structure and the initiation of the cracks in the structure.

Reference

[1] T. H. G. Megson, L. D. Landau, E. M. Lifshitz, S. Chen, G. D. Doolen and U. Alon 2010 Introduction to Aircraft Structural Analysis, vol. 35, no. 1.
[2] S. Bairavi 2016 "Design And Stress Analysis of Aircraft Wing Rib With Various Cut Outs," vol. 358, no. 4, pp. 511–514.
[3] A. Mostafa, R. De Breuker, Z. Gurdal and J. Hol January 2008 "Aircraft Stress Analysis and Structural Design," .
[4] P. James, D. M. Krishna and G. Kotresh April, 2009 "Finite Element Analysis of Inter Spar Ribs of Composite Wing of Light Transport Aircraft against Brazier Load,".
[5] N. Sukumar and T. Belytschko 1998 "The natural element method in solid mechanics," Int. J. Numer. Methods Eng., vol. 43, no. 5, pp. 839–887.
[6] J. Yvonnet, D. Ryckelynck, P. H. Lorong and F. Chinesta 2004 "A new extension of the natural element method for non-convex and discontinuous problems: the constrained natural element method (C-NEM)," Int. J. Numer. Methods Eng., vol. 60, no. 8, pp. 1451–1474.
[7] J. Yvonnet, D. Ryckelynck, P. Lorong and F. Chinesta 2003 "Interpolation naturelle sur les domaines non convexes par l'utilisation du diagramme de Voronoi contraint Méthode des éléments C-naturels," Rev. Eur. des Eléments, vol. 12, no. 4, pp. 487–509.
[8] L. Illoul, J. Yvonet, F. Chinesta and S. Clénet 2006 "Application of the Natural-Element Method to Model Moving Electromagnetic Devices," Construction, vol. 42, pp. 727–730.
[9] I. Alfaro, D. Bel, E. Cueto, M. Doblare and F. Chinesta 2006 "Three-dimensional simulation of aluminium extrusion by the α-shape based natural element method," Comput. Methods Appl. Mech. Eng., vol. 195, no. 33, pp. 4269–4286.
[10] J. Yvonnet, F. Chinesta, P. Lorong and D. Ryckelynck 2005 "The constrained natural element method (C-NEM) for treating thermal models involving moving interfaces," Int. J. Therm. Sci., vol. 44, no. 6, pp. 559–569. b

IOP Conf. Series: Materials Science and Engineering **234** (2017) 012019 doi:10.1088/1757-899X/234/1/012019

Research on the Integration of Bionic Geometry Modeling and Simulation of Robot Foot Based on Characteristic Curve

G He*, H Zhu, J Xu, K Gao and D Zhu

College of Mechanical & Electrical Engineering, Hohai University, Changzhou, Jiangsu, China

*hegang@hhu.edu.cn

Abstract: The bionic research of shape is an important aspect of the research on bionic robot, and its implementation cannot be separated from the shape modeling and numerical simulation of the bionic object, which is tedious and time-consuming. In order to improve the efficiency of shape bionic design, the feet of animals living in soft soil and swamp environment are taken as bionic objects, and characteristic skeleton curve, section curve, joint rotation variable, position and other parameters are used to describe the shape and position information of bionic object's sole, toes and flipper. The geometry modeling of the bionic object is established by using the parameterization of characteristic curves and variables. Based on this, the integration framework of parametric modeling and finite element modeling, dynamic analysis and post-processing of sinking process in soil is proposed in this paper. The examples of bionic ostrich foot and bionic duck foot are also given. The parametric modeling and integration technique can achieve rapid improved design based on bionic object, and it can also greatly improve the efficiency and quality of robot foot bionic design, and has important practical significance to improve the level of bionic design of robot foot's shape and structure.

1. Introduction

The design of robot foot working in field environment is of great significance to its performance of running efficiency and anti-sinking. Many researchers have studied the bionic design problem of the foot of ostrich, camel and gecko, and based on bionic design approach, new mechanical structure has been established according to the structure of the animal's feet, of which Ji [1-2] has designed walking mechanism according to the structural characteristic of camel's foot, and Yang [3] has designed the wheel of mars rover based on plantar morphological characters of ostrich's foot. The geometric bionics is an important content in bionic research. At present, the main idea of bionic research is to make a sample similar to the bionic object after measurement and improve the prototype after physical experiment [4-9]. But the process needs much time and high cost, and it is difficult to obtain the optimal scheme of bionic design. With the help of numerical simulation technology, finite element method can be used to replace experiment in some extent. One can reduce the time of experiment, and reduce the cost and shorten design cycle by simulation. However, because the shape of bionic object is usually very complex, the process of geometric modeling and mesh generation is tedious, which restrict the important of the efficiency and quality of bionic design.

Aiming at the deficiency of existing bionic design method, the bird's feet living in soft soil and swamp environment will be taken as researching object, and considering the flexibility of design, and the efficiency of modeling and simulation, a parametric model, which is mainly made up of characteristic of skeleton and section curves, is established with ABAQUS and the integration of geometric modeling, finite element simulation and optimization is implemented to improve the efficiency of bionic design.

2. Integrated framework of bionic geometric modeling and simulation

Bionic foot's geometric modeling is the process of building a geometric model that can represent the structural characteristic of animal's foot similar to the structure of animal's foot in shape. Figure 1 shows the integrated framework of bionic geometric modeling and numerical simulation. The structural characteristics of animal's foot are analyzed in order to find the most suitable characteristic curves and variables that can be used to define the shape of foot, toes and flippers, and then the model made up of curve network is generate to describe the shape features of foot. After creating surface model based on foot's curve network, solid model is generated and finite element meshes of foot and soil are built. In order to deal with larger deformation problem, Multi-material Arbitrary Langrage Euler method is used to describe the displacement field of soil. Then the displacement boundary conditions of soil and foot are imposed, and soil's gravity force, foot's load-bearing capacity is applied, and contact parameters between soil and foot are also defined. The results are outputted after solving the governing dynamics equations, and compare the results from different parameters after post processing; if the new results is not better than the results from the parameters used before, they should be adjusted, and then generate a new foot model with the new parameters. Repeating the above process until the best results that can be accepted are found. With the framework described in figure1, the integration of bionic geometric modeling and simulation can be implemented, and the optimization can be realized automatically. The most important part in the integrated framework is to define the characteristic curves of each part accurately, which can create more reasonable models and lay the foundation for the follow-up process. In this investigation the integrated framework is implemented in Abaqus with Python language.

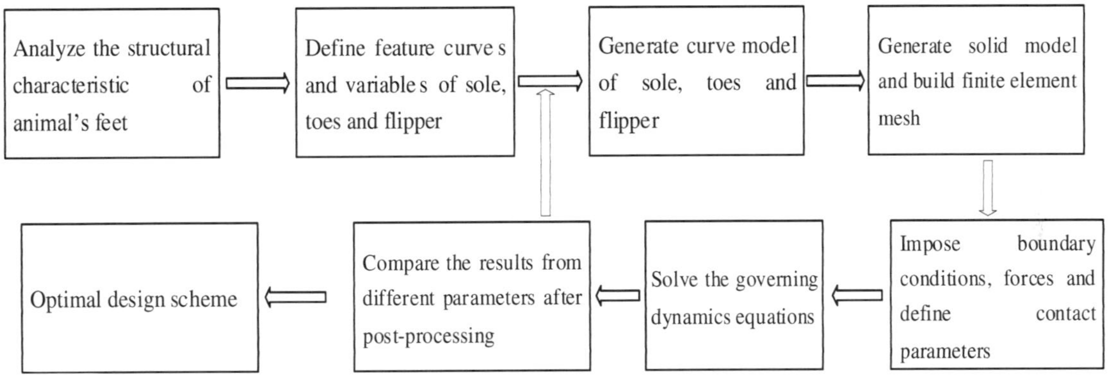

Figure 1. Integrated framework of integration of bionic foot's geometric modeling and sinking simulation.

3. Geometric model of bionic foot

There are many different species living in diverse environments, for example there are crocodiles and cranes in the swamp, ducks and frogs in soft soil, and camels, ostrich and other animals in the desert. They are evolved over long periods of time in nature and have strong abilities to adapt to various terrains. Nowadays, many domestic and foreign scholars have studied the structures of animal's feet [10-12], and one of them has applied bionic ideas to the research of anti-sinking. Many animal's feet is a very helpful to inspire the research of anti-sinking, of which ostrich and duck is very representative. Their feet structural features are mainly supported by toes that are connected together by flippers. Therefore, this paper chooses these two representative animals' feet to demonstrate ours bionic geometric modeling method, and use the modeling into the integrated framework.

3.1. Geometric modeling principle of bionic foot

The geometric models of the bionic feet are created according to the similar structural features of animal's foot. The bionic foot geometry model mainly includes three parts, which include sole, toes and flippers. The geometric model of the bionic foot is shown in figure2. The detailed modeling steps are as follows: firstly, the foot section curve S_0 is defined, which can be used to generate the sole of the foot; creating the toe skeleton curves f_s which is mainly used to define the shape of phalanxes and control the rotation angle between toes. Secondly, different toe section curves S_1, S_2, S_3 should be defined according to the skeleton curves. At the same time, the rotation angle variable β between toes and the rotation angle variable α at joints need to be defined. Thirdly, the model of flipper whose characteristic curve f_c is sketched on the plane that parallels to sole plane and the thickness h of flipper also is defined at the same time.

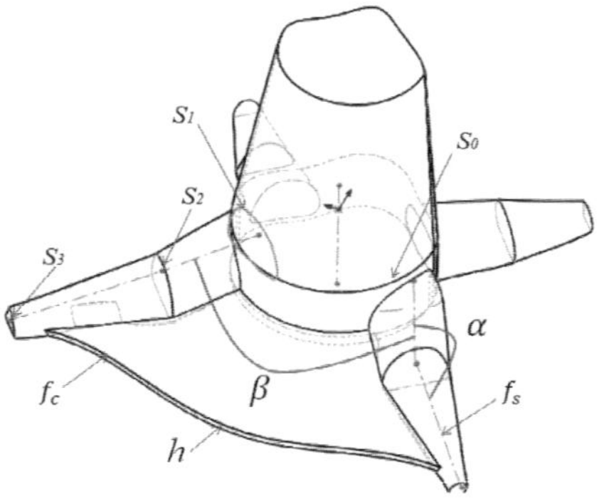

Figure 2. Conceptual model of bionic foot.

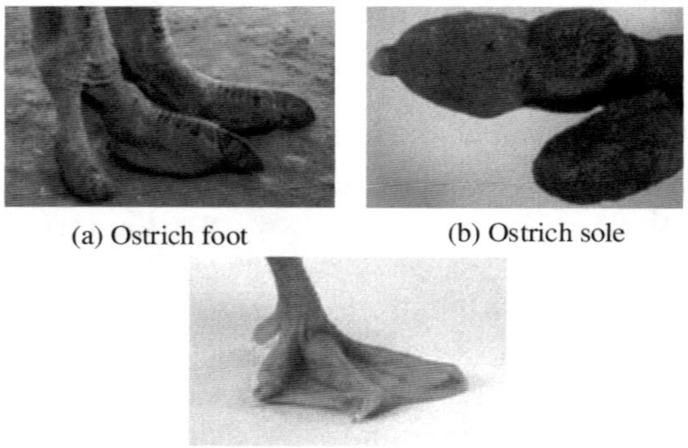

(a) Ostrich foot (b) Ostrich sole

(c) Duck foot

Figure 3. Ostrich foot and duck foot.

3.2. Structure characteristics of ostrich foot and duck foot.

Ostrich foot and duck foot are shown in figure 3, of which ostrich foot shown in figure 3a and 3b has only two toes [13], and they are third toe and fourth toe. The third toe is relatively developed, and the fourth toe plays a role of supporting the body when walking. The third toe and fourth toe are connected with a flipper. Because the ostrich lives in desert and wilderness areas, the soles of the feet, and the third toe, often contact with the sand. After a long time evolution, its sole has formed a unique plantar surface. As figure3c shown, duck foot has four toes, three of which are relatively long. The three longer toes are connected with flippers. The shorter toe plays a role of balance and support.

3.2.1. Geometric modeling of bionic ostrich foot. The toes of ostrich directly contact with sand, so the shape and structure of ostrich toe is important to the sinking property and walking efficiency of ostrich. The definition of ostrich's toes are based on its characteristic curves and section curves, and these curves are defined by a series of interpolation points or control points. The surface of toes can be built by sweeping the section curves S_i (i =1, 2, 3…) of toe along its skeleton curve f_s. The two toes of ostrich's feet do not have phalanx, and the modeling method of the third and fourth toes are the same.

Figure 4(a) gives the shape of feature curves of the third toe, which can be defined by the coordinates of fourteen points from A to N. The y coordinates of the lowest points G and J together with the highest point H in the middle groove are taken as basic parameters, which are denoted by k_1, k_2 and q respectively. The curve describing the section of bionic ostrich toe is defined in Abaqus with Python, the code is as follows:

```
s.Spline(points=((-40.5, 0.0), (-39.25, 2.5), (-35.5, 3.9), (-31.75, k1-2.75), (-24.25, k1-0.25), (-16.0,
k1-0.4), (-8.0, k1), (0, q), (8.0, k2-0.5), (16.0, k2), (24.0, k2-1.0), (32.0, k2-2.5), (37.0, k2-6.25),
(38.0, 0.0)))
```

The shape of toe's section curve can be modified by changing the value of k_1, k_2 and q. Combined with section curves, the bottom surface of bionic ostrich toe can be created by sweep method. While we give a simple model that is built by directly revolving the feature curve around the axis. The corresponding Python code is as follows:

```
s.Line(point1=(-40.5, 0.0), point2=(38.0, 0.0))
s.VerticalConstraint(entity=g[4], addUndoState=False)
p=mdb.models['Model-1'].Part(name='Part-dajiaozhi',dimensionality=THREE_D, type =
DEFORMABLE_BODY)
p = mdb.models['Model-1'].parts['Part-dajiaozhi']
p.BaseSolidRevolve(sketch=s, angle=180.0, flipRevolveDirection=OFF)
a.rotate(instanceList=('Part-xiaojiaozhi-1', ), axisPoint=(0.0, y1, 0.0), axisDirection=(0.0, 0.0, z1),
angle= β )
```

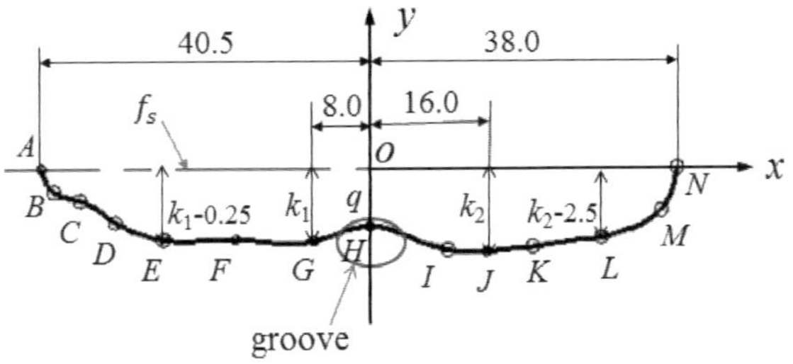

(a)

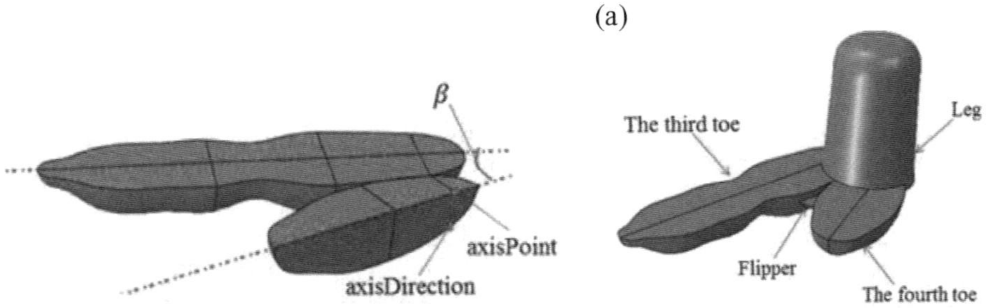

(b) (c)

Figure 4. Geometric models of ostrich.

Where β is the rotation angle variable that is used to define the relative posture between two toes' skeleton curves. When β equals -25 °, and y_1 equals -12.50mm, and z_1 equals 6.91mm, the corresponding toe model is generated as figure 4(b) shown. After creating the flipper, the sole, toes and flipper should be assembled in Abaqus. The whole structure of bionic ostrich foot is shown in figure 4(c).

3.2.2. Geometric modeling of bionic duck foot. The modeling of bionic duck's foot is simpler than that of ostrich's foot, and the geometric modeling steps of bionic duck's foot is similar to that of ostrich's foot. Each toe of bionic duck's foot is divided into three phalanxes, and there is no parameter defined between the second phalanx and the third phalanx. Figure 5(a) shows the toes skeleton curve of duck's foot and the shape of toes of the two-dimensional shape.

IOP Conf. Series: Materials Science and Engineering **234** (2017) 012019 doi:10.1088/1757-899X/234/1/012019

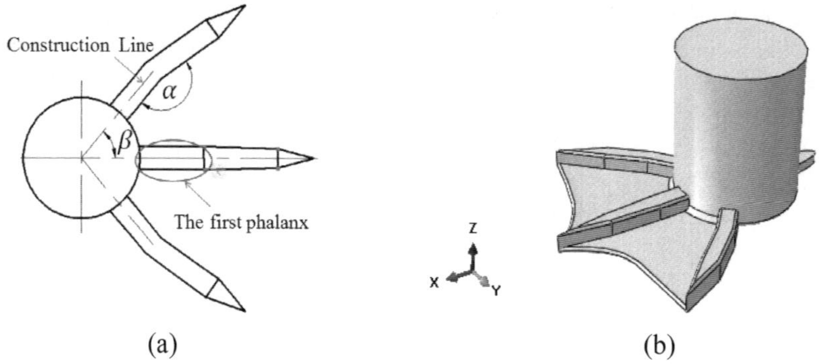

Figure 5. Geometric models of ostrich.

The following code segment only shows the rotation angle variable β of toes and the rotation variable α at toe's joint.

```
s1.ConstructionLine(point1=(0.0, 0.0), point2=(2.21, 0.0))
s1.HorizontalConstraint(entity=g[3], addUndoState=False)
```

Three construction lines should be created by using above code, which control the rotation angle of three toes. So the rotation angle β of three toes will be changed by adjusting the rotation angle between construction lines.

```
s1.AngularDimension(line1=g[4],line2=g[3],textPoint=(x1, y2), value= β ))
s.AngularDimension(line1=g[19],line2=g[10],textPoint=(x2, y3), value=162.0)
s.delete(objectList=(d[0], ))
s.AngularDimension(line1=g[19],line2=g[10],textPoint=(x3, y4), value= α )
```

The rotation of toes can be controlled by changing the rotation angle β, and the rotation of phalanxes can be controlled by changing the rotation angle α. It should be noted that the fixed constraints should be defined between the second phalanx and the third phalanx. Bionic duck foot model is shown in figure 5(b).

4. Simulation results and discussion

4.1. Parametric foot with different parameters

The bottom surface of the toe is directly contact with soil, so the toe's structure is very important to the study of anti-sinking. The parameters are used to establish different toe models are q, k_1 and k_2, h_1 and h_2, of which k_1 and k_2 are used to control the third toe's shape curve, h_1 and h_2 are used to define the fourth toe's shape curve. Two different values are taken for each parameter, which are given in Table 1. After the parameters are modified, the parameterized code is directly imported into Abaqus, and the models from two groups of parameter are generated automatically as figure 6 shown.

Table 1. Different values of each parameter.

Parameter	q/(mm)	k_1/(mm)	k_2/[mm]	h_1/[mm]	h_2/[mm]	β/[°]
Group 1	6	9.5	9.5	9.0	16.5	-35
Group 2	8	8.0	10.5	11.0	19.0	-15

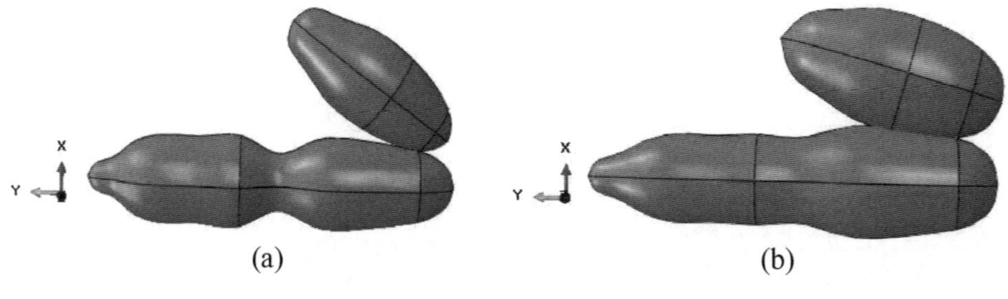

IOP Conf. Series: Materials Science and Engineering **234** (2017) 012019 doi:10.1088/1757-899X/234/1/012019

Figure 6. Parametric foot with different parameters.

Parametric method can be used to build different models by quickly adjusting the model's parameters. This lays the foundation for the simulation and optimization of the foot structure, and it is easy to realize the integration of modeling and simulation of the bionic foot in the environment of Abaqus with Python language. It should be noted that the absolute changing amount of the parameters should not be too large, otherwise it will lead to a model with impractical dimensions.

4.2. The influences of parameters k_1, k_2 and q.

In order to study the influence of the values of k_1 and k_2 on the sinking property of bionic foot, the results of displacement and velocity of soil under the bionic foot are analyzed in this section. Figure 7 shows the displacement of soil when k_1 and k_2 are 9.0 and 10.0 respectively. The maximum displacement of soil is locating at the lowest point at the bionic foot's bottom surface, and at the center of the groove, the displacement of soil is relatively lower. Figure 8 gives the velocities of soil under different values of k_1 and k_2. We can find that the maximum value of soil's velocity increases with both the values of k_1 and k_2, but the differences on the average velocities of soil under the bottom of bionic foot are small, and the whole distribution of the soil's velocities are similar.

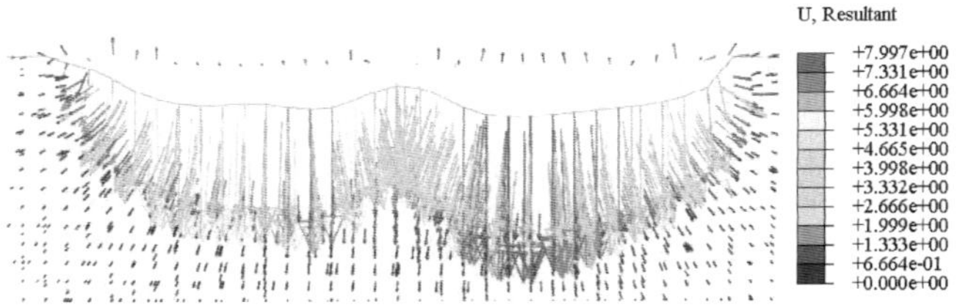

Figure 7. Displacement of soil under the foot.

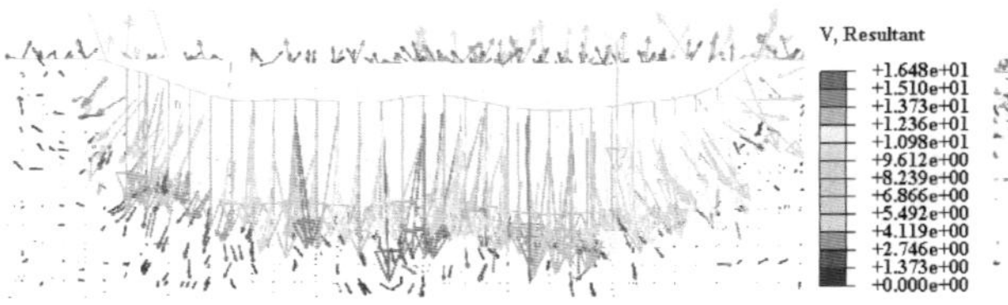

(a) k_1= 8.0, k_2= 9.0

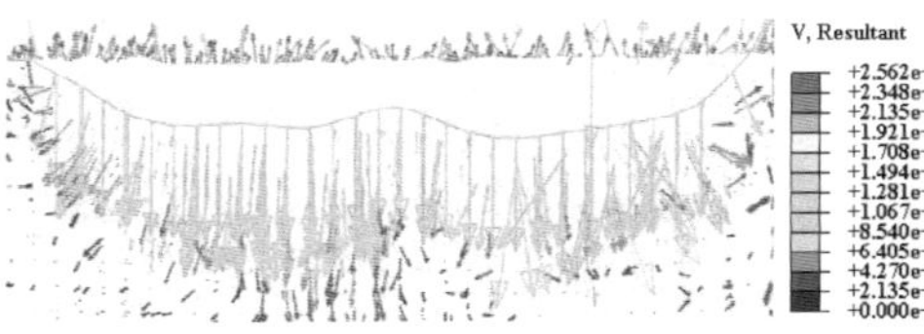

(b) k_1= 9.0, k_2= 10.0

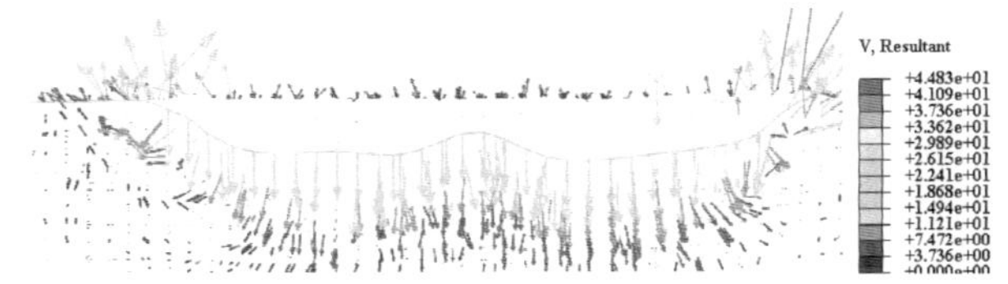

(c) k_1= 10.0, k_2= 11.0

Figure 8. The velocities of soil with different value of k_1 and k_2.

The sinking characteristics of the bionic ostrich foots with different groove dimension in the toe section curve S_7 is compared by changing the value of q from 6.0 to 8.0. Three toe models are built by Python code to simulate the sinking process, and the results force and sinkage are compared in figure 9. It can be seen from figure 9 that the overall trends of three curves are the same, but under the same force, when q is 6mm, the sinkage is the minimum.

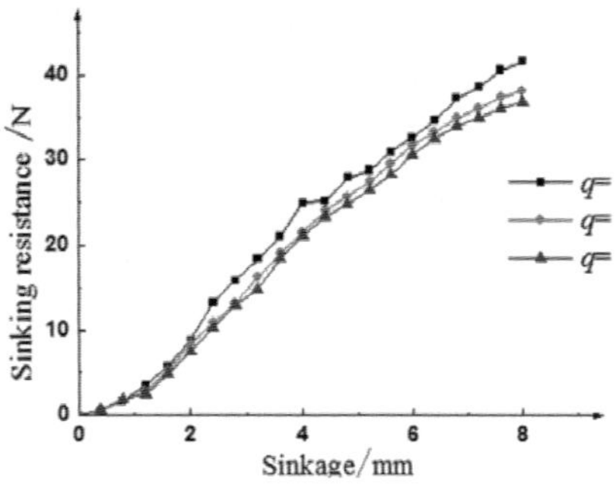

Figure 9. Sinking characteristics at different q.

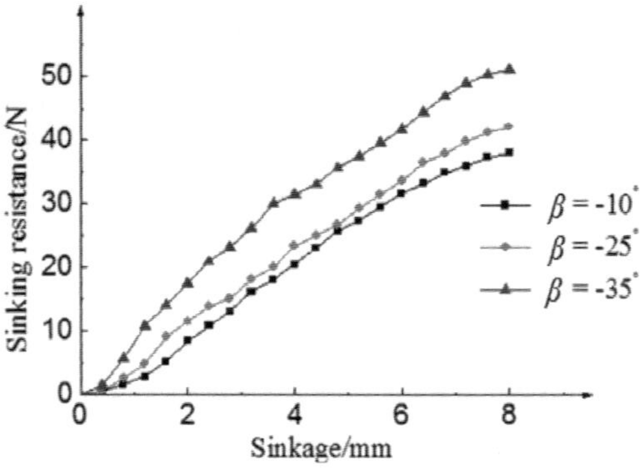

Figure 10. Sinking curves with different β.

4.3. Influence of parameter $\boldsymbol{\beta}$.

Figure 10 compares the sinking curves of the bionic ostrich foots when β is assigned the value of -10°, -25° and -35° respectively, which shows that when β equals -35° the sinkage is the smallest under the same sinking resistance. Figure 11 gives the stress distributions of soil with different values of β, and it can be found that when β is -10° , the maximum stress are smaller than that of other two angles, but the overall stress distribution is similar. If more values of β is assigned, we can reach a more meaningful result for the reference of the design of the bionic ostrich foot after comparing the results carefully.

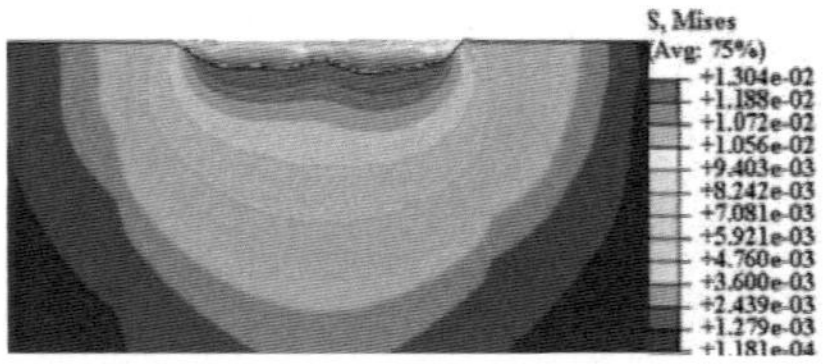

(a) β= -10°

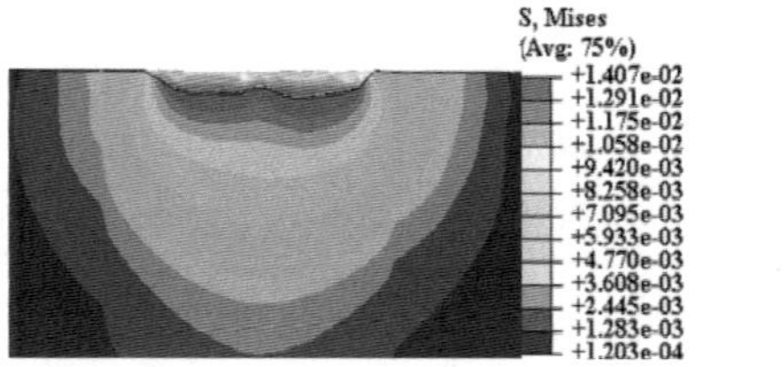

(b) β= -25°

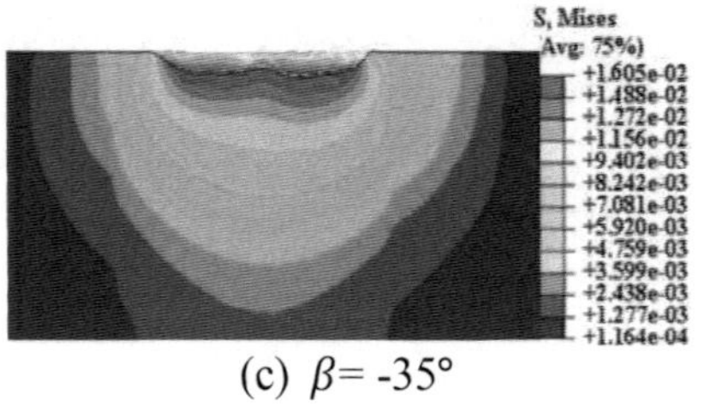

(c) β= -35°

Figure 11. Soil's stress distribution with different values of β.

5. Conclusion

In this paper, the feet of animals living in soft soil and swamp environment are taken as bionic objects. Characteristic skeleton curve, section curve, joint rotation variable, position and other parameters are used to describe the shape and position information of bionic object's sole, toes and flipper. The geometric modeling of bionic object is created by using the parameterization of the characteristic and section curves and variables. Based on this, the integration of parametric modeling and finite element modeling, dynamic analysis of sinking in soil and post-processing of bionic ostrich foot and bionic duck foot is implemented, and the detailed integration framework of the modeling of bionic geometric model and simulation is given. The bionic designs of ostrich and duck's feet are also introduced as an example, and the simulate results show that our method of parametric modeling and simulation is feasible and efficient, which can achieve rapid improvement design based on geometric bionic methods. In order to improve the optimization efficiency of bionic design, in the next step we will consider make ABAQUS integrate into the framework of Isight.

Acknowledgment
This work was supported by National Natural Science Foundation of China (No. 51375141, No. 51375140). This financial support is gratefully acknowledged.

References

[1] Ji X, Yang Z, Xu P and Chen W 1997 *J. Automotive Technol.* **4** 13-17
[2] Li J, Zhuang J, Qiu X and Wei D 1999 *Chin. J. Mech. Eng.* **10** 676-79
[3] Yang M 2015 *Bionic research in mars rover wheel based on plantar morphological characters of ostrich foot* (Jilin: Jilin University)
[4] Li J, Zhuang J and Wei D 1999 *J. Trans. CSAE*, **15** 32-36
[5] Cui Z and Yang R 2008 *Chin. J. Mech. Eng.* **44** 234-237
[6] Zhang H, Zhang Y, Wang X and Zhou S 2007 *Chin. J. of agric. Machinery* **38** 35-38
[7] Yang X, Hu H, Wang P and Zhao K 2006 *J. Mech. Eng. & Autom.* **2** 78-80
[8] Lu S, Guo C, Dai Z 2012 *J. Chin. Sci. Bull.* **26** 2463-68
[9] Yu Z, Li H, Zhang X, Zhang H and Dai Z 2011 *Chin. J. Mech. Eng.* **21** 7-13
[10] Li B, Wang X, Yu S, Sun S and Mei T 2011 *Chin. J. Mech. Eng.* **8** 897-900
[11] Menon C and Sitti M A 2006 *J. Bionic Eng.* **3** 115-125
[12] Li Y, Ge W and Fan C 2012 *J. Machinery des. & Manuf.* **5** 187-189
[13] Zhang R, Yang M, Liu H, Zeng G, Pan R and Li J 2015 *J. Jilin University* **45** 508-515

IOP Conf. Series: Materials Science and Engineering **234** (2017) 012020 doi:10.1088/1757-899X/234/1/012020

Research on the Bionics Design of Automobile Styling Based on the Form Gene

Zhao Aili[1] and Jiang Long[2]

[1] Department of Industrial Design, Heilongjiang Bayi Agricultural University, Daqing ,163000,CN
[2] Department of Product Design, Panzhihua University, Panzhihua,617000,CN

aili_fiona@126.com

Abstract. From the heritage of form gene point of view, this thesis has analyzed the gene make-up, cultural inheritance and aesthetic features in the evolution and development of forms of brand automobiles and proposed the bionic design concept and methods in the automobile styling design. And this innovative method must be based on the form gene, and the consistency and combination of form element must be maintained during the design. Taking the design of Maserati as an example, the thesis will show you the design method and philosophy in the aspects of form gene expression and bionic design innovation for the future automobile styling.

1. Form Gene and Orientation Guidance of Automobile Design

Product form gene is an expression of the hereditary knowledge about the basic vocabulary of a product and the rules of its shaping rules[1]. It is the most basic genetic element of concrete expression and constraint of object shape. And the form gene of automobile design is the combination of technology and art, which is based on the extraction of brand genes. Its special characteristics have oriented guidance to the construction of brand culture and consumer sentiment. The form gene of automobile design has similarity with the construction of its brand value. The exploration of the automobile form gene will not only enable us to trace the source and form good cultural accumulation, but also focus on the future and establish a good sustainable development context. The research and exploration of the automobile form design genes can help designers and design enterprises to accumulate and innovate their design and form unique form characteristics under the same cultural background and evolution rule. Automobile form gene is the root of the design, and the law of development is the basis for the design of adaptation, rational cognition and perceptual knowledge formed on the basis of it will be the strong guidance for the next design part. In the design, through the parts of arrangement, analysis and excavation, reconstruction and innovation of automobile shape modeling are achieved, and the final practical results will continue to be integrated into the gene chain and become an important part of the law of development. (Fig. 1)

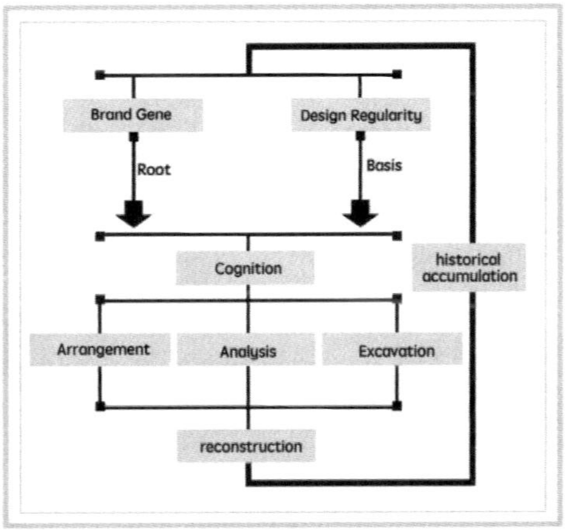

Figure 1. Automobile Modeling Gene Refining Procedure

The extraction of form characteristics of automobile brand is the arrangement and excavation of brand itself, which is also the process of self-cognition of the comprehensive system and used to accumulate brand identity elements and construct brand identity image. In 1991, professor David A. Aaker proposed the concept of brand identity, he believed that the essence of brand identity is a series of brand associations which enterprises want to form, and he expressed brand's permission and commitment for consumers[2]. Brand form gene is a collection of brand culture in the context of healthy development circulation of enterprises. It is a complex brand identity system. We will continue to design and innovate the enterprise's form genes by collecting and arranging business process assets, trend analysis and concept screening. Besides, we make the design and innovation establish emotional ties with consumers.

2. Bionics design concept and breakthrough attempt of automobile modeling

Bionics design is the external form of the existence of objects and natural substances and their symbolic meaning, and applies them to the design through the corresponding artistic treatment methods[3]. The development space of bionics design is very huge because of its infinite research objects. We carry on analysis and understanding of "goodwill" through the "revelation" of nature and use the methods and thinking of combining art with science. From the angle of humanization , we use biological structure and functional principles to design, aiming at achieving the integration of nature and science and fusion of technology and art. In the process of automobile exterior design, designers often seek design inspiration from nature, extract the form symbols or functional models from the natural species, and convert them into a practical and intimate form. This kind of natural imitation of products is in line with people's natural association and value orientation, which is a design concept with good experience and good external form.

In recent years, there have been breakthroughs of exploration and attempts in the bionics design of automobile shape modeling, aiming at enhancing the aesthetic degree of design and meanwhile increasing the emotional emotions, visual effects, characteristics of the times and interactive experience conveyed by automobile shape modeling. For instance, in the book of *My world is Round* written by Luji · Colani, a famous German industrial design master, he summed up his design ideas that, "when we talk about bionics design, we need to accept a simple fact that the structure of the spider web has much more unbelievable advantages than the large number of structures made by mankind, and we should be able to find solutions from the advantages of nature" [4]. His design emphasizes the integration of man and nature and thinks that the beauty of design is reflected in the return to nature and the product should serve the people[5]. His design made for Ferrari, Porsche, Mercedes-Benz, BMW and other companies well interpreted the concept and connotation of bionics

design. (Fig. 2-3)

Figure 2. BMW design

Figure 3. Ferrari design

3. Bionics design trends and design errors of automobile shape modeling

In the current study, bionics design is generally divided into form bionics, texture bionics, body bionics, structure bionics, function bionics, color bionics and intention bionics[6]. Therefore, in the future design study, we can extend from the above aspects, and use the design method of" insight - analysis - divergence - reconstruction", aiming at seeking for the best balance between nature and design. Insight and analysis can help us establish a good cognitive system and form a clear design research context. The divergent way of thinking can expand the design ideas and let us re-examine the connection and link between natural elements and automobile exterior shape. Finally, we finish the final design through reframing way and achieve good design results. In this process, we can well integrate the concept of bionics design into automobile modeling design in the flow of "understanding - innovation - achievement".

However, some automobile companies still have some design errors in the design and development process: they blindly pursue innovation and only consider the modeling divergence way of this automobile and automobiles with same brand. Although, new automobiles have wholly new modeling forms, they lack inheritance of brand gene. Therefore, in the design process, one the one hand, we achieve bionics design by drawing inspiration from nature. On the other hand, we comb the brand gene context of enterprises, so that the final design results integrate and reconstruct original characteristic and bionic elements and achieve true evolution of exterior modeling design.

4. Automobile modeling bionics design flow, methods and application cases based on form genes

Bionics design inherits the corporate brand genes and includes the following steps(Fig. 4): firstly, the brand cultural materials are summarized to form from genes (shape features, traditional colors and cultural contexts). And we extract bionic elements (image form, color semantics , cultural representation, etc.) from natural objects. And then, we integrate and transform form genes and bionic symbols by using divergent and subversive thinking way. We carry out design positioning, determine design keywords, draw scheme sketches, achieve innovative design results, perfect the arrangement and form the final exterior design modeling form. The whole process fully embodies the effective combination of design genes and bionic concepts, and the design should not be separated from the actual design ideas and methods.

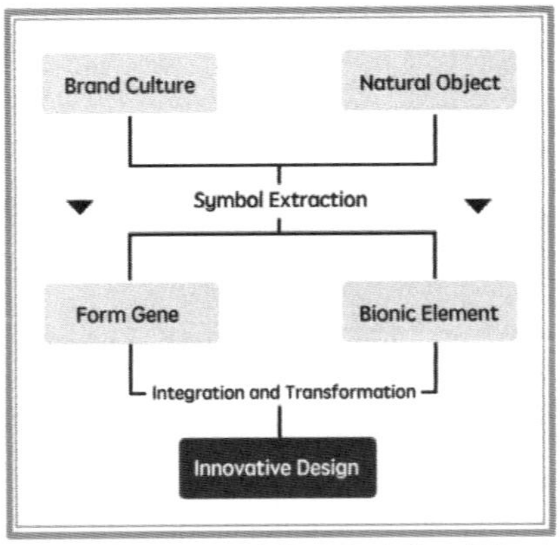

Figure4. Automobile Modeling Bionics Design Flow Based on Form Gene

Figure5. Bionic prototype - crane

We Integrates and explores form genes and bionic symbols in Maserati – cran design. Maserati is known as the classic and eternal representation of flowing sculpture, which is designed for those who seek the ultimate quality of life and pure driving pleasure and has its own unique cultural connotation. It is an indispensable part of Italy in the long and glorious history. Therefore, We focus on the analysis and exploration of existing products in the design of Maserati - cran, so as to seek for new breakthroughs.

Maserati has entered China since 2004, its fashionable and elegant atmosphere quickly got the recognition and favor of Chinese consumers. Currently, China has become the country that has the highest growth rate in the Maserati market. It can be seen that it has huge sales potential in China. However, with the awakening of traditional Chinese cultural awareness, consumers are more looking forward to the birth of the automobile shape with national consciousness. So in the design, we takes into account consumer demand and its development trend in Chinese market and ingrates Chinese traditional culture elements into the design, which not only inherited the Maserati gene semantics, but also used bionic way to achieve the Chinese traditional culture connotation.

Maserati models interpret centuries of elegance, luxury, sports and passion of the brand together. Maserati perfectly fuses luxury with unrestrained sports nature, pays equal attention to comfortable and passionate driving pleasure. Beides, the elegance and refinement conveyed by Maserati are also genes that need to be inherited in the design and important elements of design expression. Therefore, in the body design, in addition to the inheritance of the original modeling elements of Maserati - cran, it also integrates bionic elements. Maserati's iconic grille and dynamic outline modeling become the focus of achieving reconstruction of design genes. Besides, Maserati - crane fuses the image characteristics of crane.crane has a lofty position in Chinese culture, which represents the longevity, luck and elegance. And its flying speed and endurance are also quite amazing, which in in line with Maserati's concept of elegance, luxury, sports and passion(Fig.5). In the modeling design, bionics design is used. The perfect arc shape of crane from its neck to the tail is extracted and fused into the waist line design of the body. So that the body has natural mobility. The beautiful arc extends from the grille to the side of the body and then to the taillight, which is in line with the aerodynamic requirements. And the waist is more aesthetic. In addition, the wrapped form of the bottom of the body makes the whole body be surrounded by plump wings. The whole automobile shape looks more stable and heavy, and the proportion of the body also takes the shape of flying crane as a reference, which achieved flexible and fashionable flavor of the body and showed its elegant and unique aristocratic temperament(Fig.6-7).

IOP Conf. Series: Materials Science and Engineering **234** (2017) 012020 doi:10.1088/1757-899X/234/1/012020

Figure6. Front face design of Maserati – crane

Figure7. Tail design of Maserati - crane

The wheel is designed in concert with the overall body temperament. large wheels can show Maserati's eternal pursuit of passion, talent and perfection. The wheels add some embossed line decorations that are extracted from the crane's feather shape on the surface of the multiple spokes, which increased the flexible feeling of design elements. Therefore, the whole wheel does not seem too serious, yet elegant and sporty, coupled with a smooth silver effect. And the unique Maserati design genes and bionic elements are combined together(Fig.8).

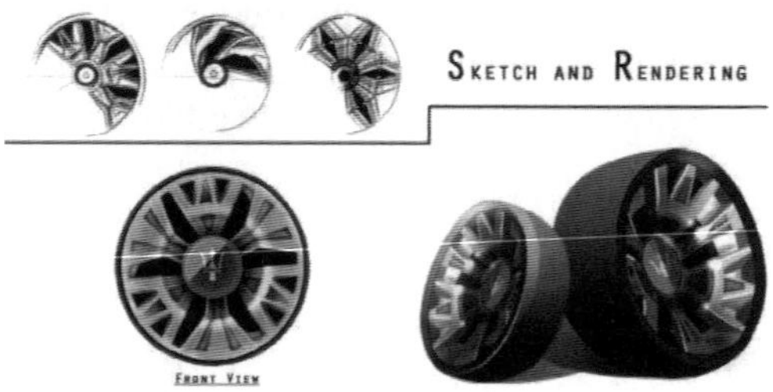

Figure8. Wheel hub design

Figure9. Automobile light design

As an important head component, headlight is the representation of the charm of the front and constitutes important facial features with other components of the front together. Headlight modeling element takes "flying and dancing crane" as the bionic prototype, which brought more dynamic visual effects for the design of automobile front face. Meanwhile, the automobile light tries a new proportional relationship, stretches the light modeling to long and narrow shape and divides it into main/subsidiary parts. Therefore, headlights effect is more discriminating, and the front face of the body is more moderate than and that of the existing sports cars. It is connected with the atmospheric grille, which effectively enhances the visual tension and looks very confident(Fig.9).
Through the practices and study of bionics design based on form design gene, the author carries out innovative fusion of Maserati brand culture gene and morphological characteristics of crane, which is an important link in the development of brand culture.

5.Conclusion

In this paper, the author studies automobile modeling bionics design based on morphological gene, summarizes the design methods and design flow of this process, and carries out practical interpretation of the Maserati - cran design plan. Through the analysis and research, the author wants to remind designers that they can not view design issues superficially in the design process, they should continue and upgrade the brand culture and use design concept of bionics to conduct breakthroughs and exploration. This kind of bionic design research that can convey the brand gene takes nature as the teacher, stimulates creative ideas and eventually acquires more innovative design results, which not only help enterprises get rid of plagiarism or design disjointed state, but also enable enterprises to start down the developing road of independent innovation.

IOP Conf. Series: Materials Science and Engineering **234** (2017) 012020 doi:10.1088/1757-899X/234/1/012020

References

[1] Sun Shouqian, Bao Enwei & Pan Yunhe 1999 China. Component Feature Model Oriented to Product Layout Design. *Computer-Aided Design & Computer Graphics*.**11**.
[2] David A. Aaker 1995. Building Strong Brands. *Big Apple Tuttle-Mori Literary Agency Inc*
[3] Chen Wei 2010 China.Bionic Design and Application of Product Form. *Packaging Engineering*.**8**.
[4] Huo Yuhua, Dai Junjie & Dong Zhaohui 2005 China. My World Is Round. *Aviation Industry Press*.
[5] Tian Jun 2013 China. Nature: Source and Direction - Luji Colani's Bionic Design *Decoration*.**4**.
[6] Ren Luquan & Liang Yunhong 2012 China. Coupling Bionics. *Science Press*.

IOP Conf. Series: Materials Science and Engineering **234** (2017) 012021 doi:10.1088/1757-899X/234/1/012021

Structural Topograohy Optimization for Thin-plate Based on Isogeometric Analysis

Zhihui Zou, Denglin Zhu*, Lilv Lou, Huishan Liu and Gang He

College of Mechanical and Electrical Engineering, Hohai University, Changzhou 213022, China

*Corresponding author's Email: cz_zhudenglin@sina.cn

Abstract: To satisfy high performance requirements for some structures, it is desirable for structure to be designed smaller, thinner and lighter in automobile and aerospace industries, etc. One prominent example is the application of topography stiffened structures. The stiffness of a structure can be increased obviously by stiffeners under a given amount of material. This paper gives an approach to optimize the structural topography design of thin-walled structures automatically based on isogeometric analysis. Optimization model, sensitivity analysis and optimization process for structural topography design are described and studied in details. The objective function is minimum compliance of structure, and the y-value(vertical coordinate value) of control points as design variables. Then structural topography is formed with changing the y-value of control points in design domain according to the specific algorithm. Several examples illustrate that the proposed method works well.

1. Introduction

Structural optimization is aimed at obtaining high performance of structures through changing the initial size, shape and topology of structure, or material properties within a series of prescribed objectives and constraints[1-5]. In automobile and aerospace etc. the optimization of stiffened plate structure has significant applications, due to stiffeners increasing small weight but the total stiffness improving obviously. However, what the best stiffentopography is and how to get the optimum size, shape and layout of structural topography are important issues.

During the past decades, the optimal design for stiffened structure has received great attention. Through optimizing thickness distribution, CHENG and OLHOFF[6] considered the problem of improving stiffness of solid elastic plates. CHUNG and LEE[7] studied the optimal shape and location of rib by topology optimization based on variable density approach. In this research, many discrete shell elements are first added to ground plate, which densities are determined afterwards by the optimization procedure to minimizing compliance of structure. Finally, the shell elements with higher densities constitute the structural rib. To get the optimized configuration of the ribs, vertical shell elements have to be located densely enough to the existing structure in advance.

LAM and SANTHIKUMAR[2] investigated the optimum of the stiffeners. In this research, element thickness is changed to achieve uniform strain energy density distribution. DING and YAMAZAKI[8; 9] suggested the growing and branching tree model to optimize stiffener patterns of plate structures. Stiffeners in the ground structure start from seeds specified in advance, and extend according to the growing and branching rules. The positions of seeds have critical influence on the final stiffener layout. So, how to preset them is full of challenge.

However, up to now, the structural topography optimization is specifically research about stiffer location based on the traditional finite element method, and rarely to material distribution of structural surface. Presented optimal model by FEM will lead to a large number of design variables to achieve the optimum stiffener shape and layout. Moreover, frequent transform between analysis model and design model is inconvenient. This paper presents a new method to deal with structural topography generation on a plate, which is based on IGA (Isogeometric Analysis). The analysis model is unified with the design model (i.e., geometry model) represented by NURBS and the corresponding control points, which avoids the difficulties in transform between these two models. Because the optimal result is described by NURBS, it can be returned to CAD system directly.

The outline of this paper is as follows: In section 2, the structural topography optimization problem is defined, optimization model and process are described, and analytical sensitivities on linear elasticity problem are formulated. In section 3, several numerical examples of structural topography optimization are presented. Finally, a summary and outlook are given in section 4.

2. Problem definition of Structural topography optimization

The goal of this paper is to generate optimum structural topography on plate automatically to increase the stiffness of plate structure. Assuming that only a prescribed amount of material can be added to the ground plate to generate structural topography. The numerical models of plate are defined using NURBS, in which, two quadratic elements with four control points are employed in the through-thickness direction and biquadratic NURBS basis functions are utilized for constructing the surface.

In the isogeometric method, analysis model and design model are represented by the NURBS, and changed with control points movement. So, this paper chose the coordinate of control points as the design variables. Since the structural topography is usually distributed on the surface of the plate, the design variables are further limited to y-value(vertical coordinate value) of the control points. As shown in Figure. 1, the control points on the lower surface are fixed, while those on the upper surface can move upward, which results in the topography growing vertically on the ground plate. To homogenize the meshes, the middle control points are always distributed evenly in the thickness direction during the optimization process. With change of y-values of the control points, the topography shape on the upper surface is changed automatically. When several adjacent control points move upward together, the stiffened plate is produced as depicted in Figure. 2. The distribution of structural topography is determined automatically by the optimization algorithm.

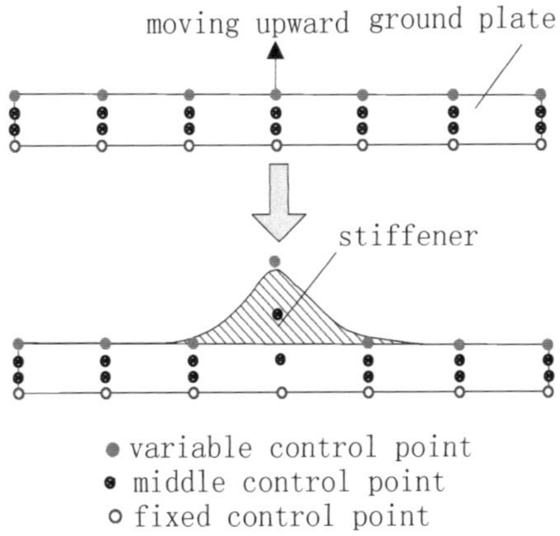

Figure 1. Cross-section change in optimization

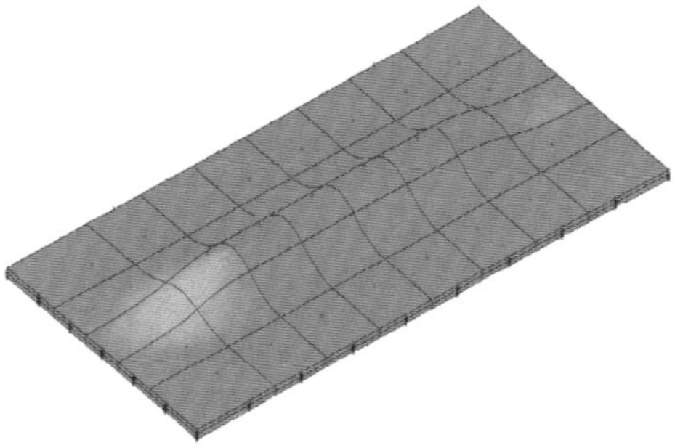

Figure 2. Structural topography plate

2.1. Optimization model

In this paper, structural topography, by adding a prescribed amount of material on plate, is optimized to minimize the compliance of the structure. As y-values of the upper control points are assigned to design variables, the optimal model is expressed as:

$$\begin{aligned}&\text{minimize} \qquad f=\mathbf{F}^{\mathrm{T}}\mathbf{u}\\&\text{subject to}\\&\begin{cases}g=V-V_{\text{limit}}\le 0\\ \underline{\varphi}_i\le\varphi_i\le\overline{\varphi}_i,\ \ i=1,\dots,n\end{cases}\end{aligned} \tag{1}$$

where $\mathbf{F}$ is the force vector, $\mathbf{u}$ is the displacement vector, g is the constraint function, V is the final volume of stiffened plate, V_{limit} is the upper limit of the final volume prescribed in advance, $s=(\varphi_1,\dots,\varphi_n)^{\mathrm{T}}$ is the vector of design variables, $\underline{\varphi}_i$ and $\overline{\varphi}_i$ are given lower and upper bounds on the design variables.

2.2. Sensitivity analysis

The gradient-based MMA [10] is used, which requires both structural response and sensitivity information. In MMA, the optimization problem is replaced by a strictly convex approximating subproblem in each step of the iterative process, which is mainly based on the sensitivity information at the current iteration points. Both WALL et al.[11] and CHO et al.[12] have formulated the analytical sensitivities for positions of control points. QIAN[13] has further investigated the full analytical sensitivity method, in which both the positions and weights of NURBS control points are served as design variables in structural shape optimization. Based on previous research work, sensitivity analysis of elasticity problem is derived in this section.

For the objective function stated in equation(1), the derivative can be calculated as

$$\frac{\mathrm{d}f}{\mathrm{d}\varphi}=\left(\frac{\mathrm{d}\mathbf{F}}{\mathrm{d}\varphi}\right)^{\mathrm{T}}\mathbf{u}+\mathbf{F}^{\mathrm{T}}\left(\frac{\mathrm{d}\mathbf{u}}{\mathrm{d}\varphi}\right) \tag{2}$$

where $\mathbf{dF}/\mathrm{d}\varphi$ can be assembled from the derivatives of the components of the local force vectors in the ith direction, $\mathbf{dF}_i^e/\mathrm{d}\varphi$, similarly with the assembling of element force vector $\mathbf{F}$. To express concisely, we simplify the notation by defining $\mathrm{d}()/\mathrm{d}\varphi=()_{,\varphi}$.

$$\begin{aligned}\mathbf{F}^e_{i\,,\varphi}&=\int_{\hat{\Omega}^e}\left(\mathbf{R}^e{}_{,\varphi}f_i\left|\mathbf{J}\right|+\mathbf{R}^e f_{i,\varphi}\left|\mathbf{J}\right|+\mathbf{R}^e f_i\left|\mathbf{J}\right|_{,\varphi}\right)\mathrm{d}\hat{\Omega}\\&+\int_{\hat{\Gamma}^e}\left(\mathbf{R}^e{}_{,\varphi}h_i\left|\mathbf{J}\right|+\mathbf{R}^e h_{i,\varphi}\left|\mathbf{J}\right|+\mathbf{R}^e h_i\left|\mathbf{J}\right|_{,\varphi}\right)\mathrm{d}\hat{\Gamma}\end{aligned} \tag{3}$$

Specifically, for a concentrated load, it can be derived as:

$$\begin{aligned}\mathbf{F}^e_{i\,,\varphi}&=\int_{\hat{\Omega}^e}\left(\mathbf{R}^e{}_{,\varphi}f_i\left|\mathbf{J}\right|+\mathbf{R}^e f_{i,\varphi}\left|\mathbf{J}\right|+\mathbf{R}^e f_i\left|\mathbf{J}\right|_{,\varphi}\right)\mathrm{d}\hat{\Omega}\\&+\mathbf{R}^e{}_{,\varphi}\left|\mathbf{J}\right|p_i+\mathbf{R}^e\left|\mathbf{J}\right|_{,\varphi}p_i\end{aligned} \tag{4}$$

The derivative $\mathbf{du}/\mathrm{d}\varphi$ can be achieved according to the following function as

$$\mathbf{K}_{,\varphi}\mathbf{u}+\mathbf{K}\mathbf{u}_{,\varphi}=\mathbf{F}_{,\varphi} \tag{5}$$

For further details, see [13]. With all the derivations stated above, the differentiation equation(2) can be solved. For the constraint function given by equation(1), the differentiation can be calculated as

$$g_{,\varphi}=V_{,\varphi}=\iiint_{\hat{\Omega}}\left|J\right|_{,\varphi}\mathrm{d}\hat{\Omega} \tag{6}$$

Where

$$|J|_{,\varphi} = |J|\mathrm{Tr}\left(J^{-1}J_{,\varphi}\right) \tag{7}$$

and

$$J_{,\varphi} = \begin{bmatrix} \frac{\partial x_{,\varphi}}{\partial \xi} & \frac{\partial y_{,\varphi}}{\partial \xi} & \frac{\partial z_{,\varphi}}{\partial \xi} \\ \frac{\partial x_{,\varphi}}{\partial \eta} & \frac{\partial y_{,\varphi}}{\partial \eta} & \frac{\partial z_{,\varphi}}{\partial \eta} \\ \frac{\partial x_{,\varphi}}{\partial \zeta} & \frac{\partial y_{,\varphi}}{\partial \zeta} & \frac{\partial z_{,\varphi}}{\partial \zeta} \end{bmatrix} \tag{8}$$

Because the design variables are the vertical coordinate value of upper control points, so

$$\begin{aligned} x_{,\varphi} &= y_{,\varphi} = 0 \\ z_{,\varphi} &= R_{i,j,k}^{p,q,r}(\xi,\eta,\zeta) \end{aligned} \tag{9}$$

Hence, both the sensitivities of objective and constraint functions required in MMA can be evaluated.

In the process of optimization, the structural topography grows or disappears with the vertical coordinate value of control points changing in prescribed ranges. The optimization process continues until the convergence criterion equation (10) is satisfied.

$$\varepsilon = \left|\frac{f^{(k)} - f^{(k-1)}}{f^{(0)}}\right| \le 5\times10^{-5} \tag{10}$$

where $f^{(k)}$ is the objective function value at the kth iteration.

3. Numerical examples

In this section, the feasibility of the proposed method will be demonstrated by three examples. The first example is a cantilever plate , the second is a completely supported square plate and the third is a semi-annular cantilever plate. In three examples, Young's modulus is E=2.1×10^5, Poisson's ratio v=0.3.

3.1. Cantilever plate

This example is a cantilever plate with dimensions l×a×h=10×7×0.1, as shown in Figure 3(a). The initial geometry is constructed from 19×10×4=760 control points and knot vectors are

$$\begin{aligned} \Xi &= \left\{0,0,0,\frac{1}{17},\frac{2}{17},\ldots,\frac{16}{17},1,1,1\right\} \\ \mathrm{H} &= \left\{0,0,0,\frac{1}{8},\frac{2}{8},\ldots,\frac{7}{8},1,1,1\right\} \\ \mathrm{Z} &= \left\{0,0,0,0.5,1,1,1\right\} \end{aligned}$$

IOP Conf. Series: Materials Science and Engineering **234** (2017) 012021 doi:10.1088/1757-899X/234/1/012021

The entire upper surface of the plate is chosen as the design domain which includes 19×10=190 control points. Note that in our examples, the optimization variables are the vertical coordinate value of upper control points, and all control weights are set to 1. In this example, the optimization variables are limited in the range of $0.1 \le \varphi_i \le 0.5$.

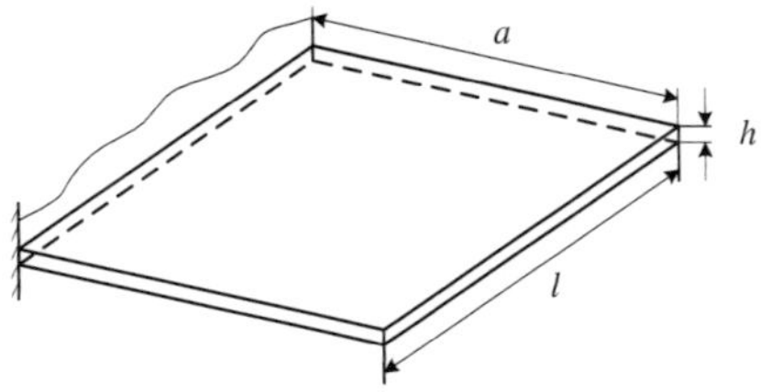

(a) Initial design

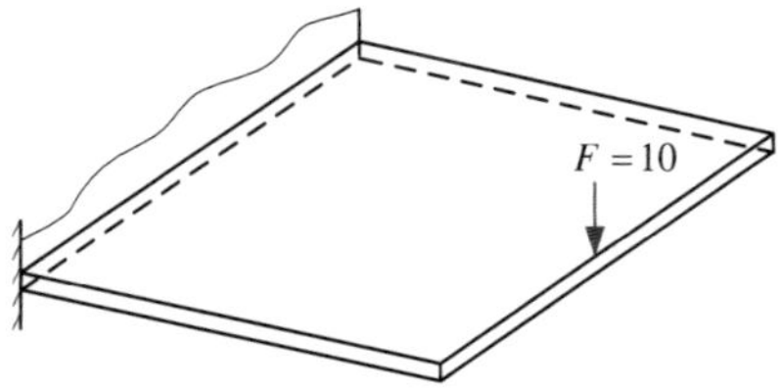

(b) One force

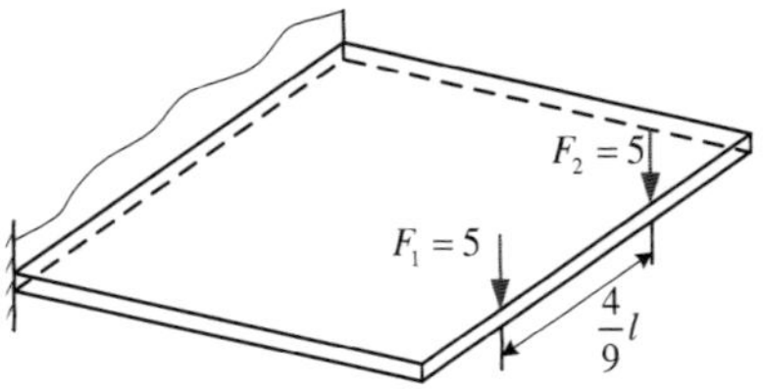

(c) Two forces

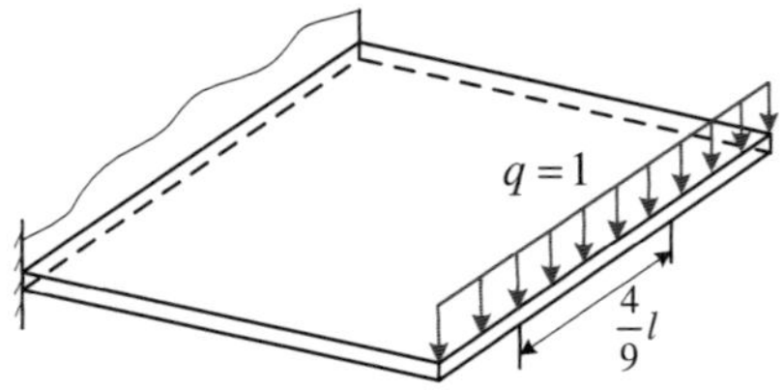

(d) Distributed force

Figure 3. Initial model of cantilever plate and loading conditions

In this example, three loading conditions are considered. In the first situation, a concentrated load $F = 10$ is applied at the midpoint of the free end, depicted in Figure 3(b). The upper limit of the final volume V_{limit} is restricted to $1.2V_0$, where V_0 denotes the original volume. In the second situation, two equal forces $F_1 = F_2 = 5$ are applied at the two sides of the free end respectively, see Figure 3(c), and V_{limit} is restricted to $1.25V_0$. In the third one, a distributed force $q = 1$ is applied as depicted in Figure 3(d), and $V_{\text{limit}} = 1.5V_0$.

The optimized results for three loading conditions are gotten after 19,18 and 20 iterations, respectively, as shown in Figure 4(a), 4(b) and 4(c). Convergence processes for both structure compliance and volume for three cases are shown in Figure 5.

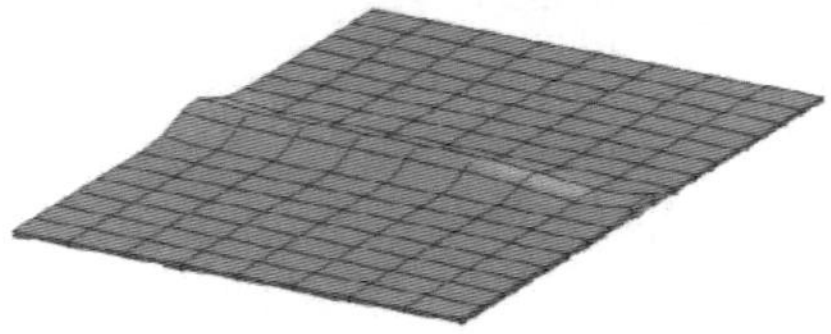

(a) Optimized design for one force

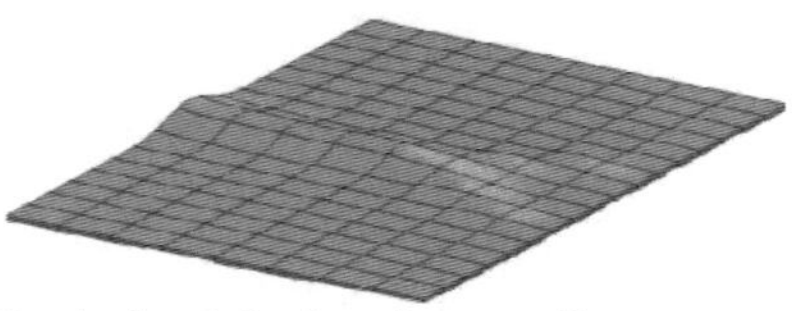

(b) Optimized design for two forces

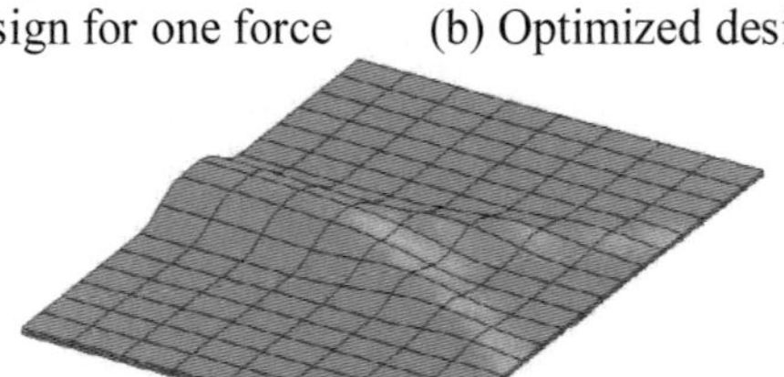

(c) Optimized design for distributed force

Figure 4. Optimized designs of cantilever plate under two loading conditions by suggested method

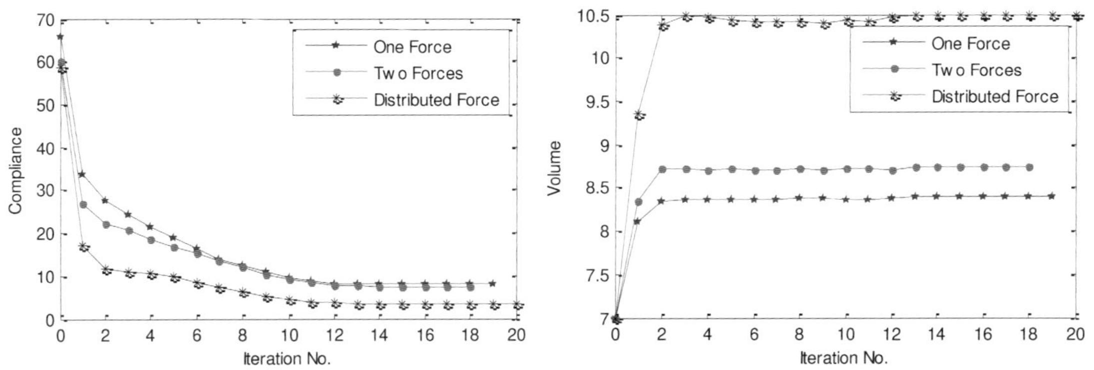

(a) Convergence process for compliance (b) Convergence process for volume

Figure 5. Convergence processes of structure compliance and volume for three cases

3.2. Clamped square plate

This example is a square plate with four edges clamped and a concentrated force applied at the center point of the upper surface, as shown in Figure 6. The ratio of length to thickness of the initial clamped square plate is 100.

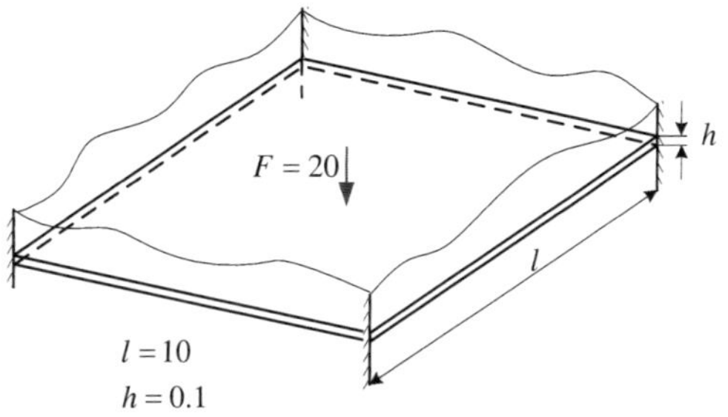

Figure 6. Initial model of clamped square plate and loading condition

The NURBS solid is defined by the following open knot vectors and the corresponding $13\times13\times4$=676 control points. All control weights are set to 1, and the NURBS solid degenerates to a B-spline solid.

Analogously, the vertical coordinate value of the upper 169 control points are chosen as the optimization variables. The upper limit of the final volume V_{limit} is restricted to $1.4V_0$ and the allowable vertical positions of the control points are restricted in the range of $0.1 \le \varphi_i \le 0.5$. The optimized solution is gotten after 19 iterations, as shown in Figure 7(a). The convergence processes of the structure compliance and volume are depicted in Figure 8.

$$\Xi = \left\{0,0,0,\frac{1}{11},\frac{2}{11}...,\frac{10}{11},1,1,1\right\}$$

$$\mathrm{H} = \left\{0,0,0,\frac{1}{11},\frac{2}{11}...,\frac{10}{11},1,1,1\right\}$$

$$\mathrm{Z} = \{0,0,0,0.5,1,1,1\}$$

The result of topology optimization of square plate with four edges clamped has been given by CHUNG and LEE[7] using variable density method. The topology optimal result is shown as in Figure 7(b), which is also similar to the distribution of structural topography in Figure 7(a).

Although the topology optimization method by reducing material is conversely with the proposed method by adding materials, both of them can obtained the similar layout of material. However, the suggested method has several advantages. Firstly, a distinct distribution of structural topography can be gotten by using the suggested method. Secondly, the computational cost of the suggested method is relatively lower, as the plate does not need to be discretized into a very large number of elements, as in the topology optimization method.

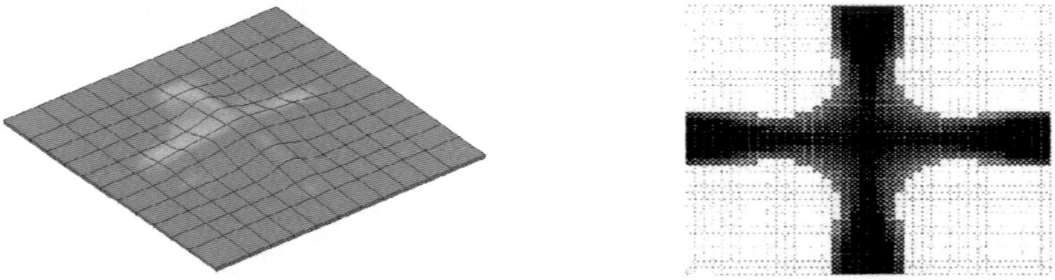

（a） Result from suggested method (b) Result from variable density method by CHUNG

Figure 7. Optimized design of completely supported square plate

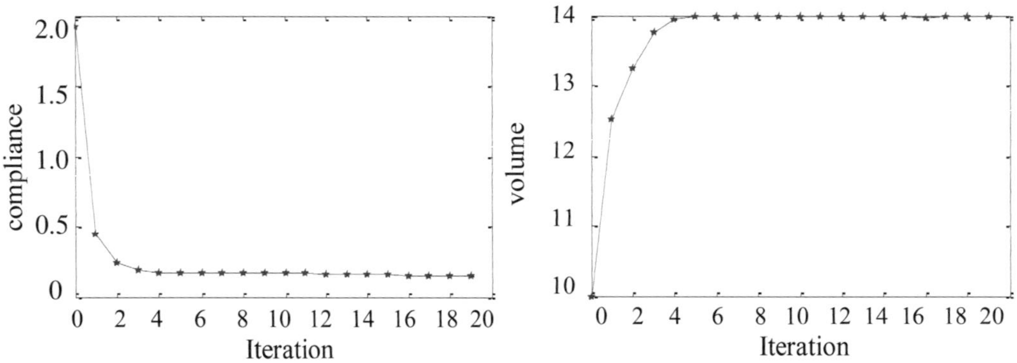

(a) Convergence process for compliance (b) Convergence process for volume

Figure 8. Convergence process for compliance and volume

3.3. Semi-annular cantilever plate

This example is a semi-annular plate with its two ends clamped and a distributed force applied along the outer edge of the upper surface, as illustrated in Figure 9. The initial geometry is build from 22×14×4=1232 control points and the corresponding knot vectors

$$\Xi = \left\{0,0,0,\frac{1}{18},\frac{2}{18}...,\frac{9}{18},\frac{9}{18},\frac{9}{18}...,\frac{16}{18},\frac{17}{18},1,1,1\right\}$$

$$\mathrm{H} = \left\{0,0,0,\frac{1}{12},\frac{2}{12}...,\frac{11}{12},1,1,1\right\}$$

$$\mathrm{Z} = \{0,0,0,0.5,1,1,1\}$$

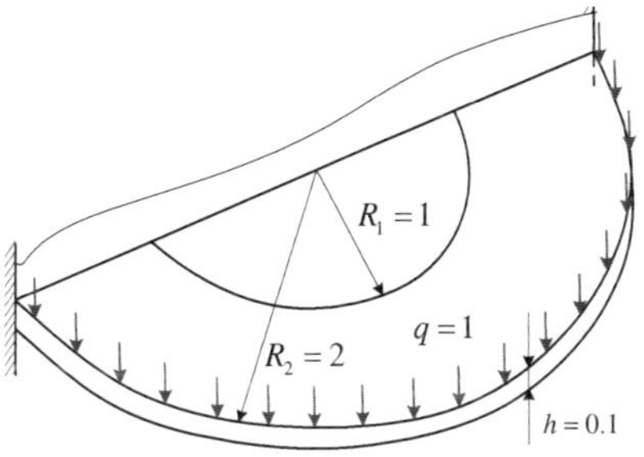

Figure 9. Initial boundary condition of semi-annular cantilever

As the preceding examples, the vertical coordinates of upper control points are chosen as the design variables. The upper limit of the final volume V_{limit} equals to $1.2V_0$ and the permissible range of design variables is limited to $0.1 \le \varphi_i \le 0.3$. The optimized result is reached in 19 iterations, as depicted in Figure 10. The convergence of both compliance and volume are shown in Figure 11.

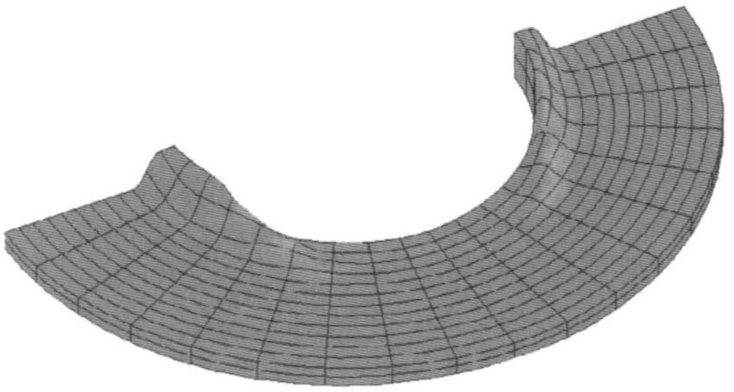

Figure 10. Optimized design of semi-annular cantilever plate

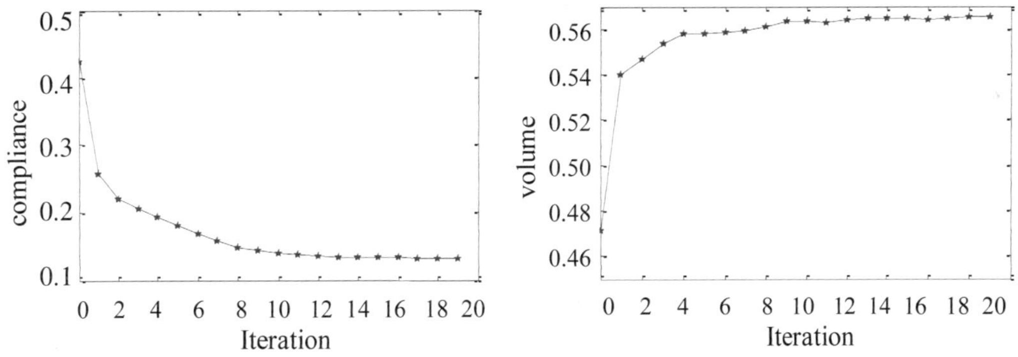

(a) Convergence process for compliance (b) Convergence process for volum

Figure 11. Convergence process for compliance and volume

4. Conclusions

In this paper, a new structural topography optimization technique based on isogeometric analysis is proposed. Since the analysis model is described by NURBS, this method can generate smooth and continuous structural topography without any post-processing because of modeling based on IGA. The integration of design and analysis model based on IGA eliminates the connecting problem between both analysis model and CAD model. So, the optimized structure can be directly delivered to the CAD system. This method can also be used to improve the dynamic characteristics of structure, such as natural frequency and frequency response function, etc.

Acknowledgement
This work is supported by the National Natural Science Foundation of China (Grant No. 51375140, 51375141) and the Fundamental Research Funds for the Central Universities (Grant No. BZX/12B101-21 and 2013B18214).

The author would like to thank Dr. Cosmin Anitescu from Bauhaus University-Weimar for his valuable suggestions.

References
[1] Haftka RT, Grandhi RV 1986 Sructural shape optimization - a survey. Computer Methods in Applied Mechanics and Engineering 57 (1):91-106. doi:10.1016/0045-7825(86)90072-1
[2] Lam YC, Santhikumar S 2003 Automated rib location and optimization for plate structures Structural and Multidisciplinary Optimization 25 (1):35-45. doi:10.1007/s00158-002-0270-7
[3] Morris A 1982 Foundations of structural optimization: a unified approach John Wiley & Sons, New York
[4] Olhoff N, Taylor JE 1983 On structural optimization. Journal of Applied Mechanics, Transactions ASME 50 (4 b):1139-1151
[5] Rozvany G, Kirsch U 1995 Layout optimization of structures Applied Mechanics Reviews 48:41-111
[6] Cheng KT, Olhoff N 1981 Investigation concerning optimal design of solid elastic plates International Journal of Solids and Structures 17 (3):305-323
[7] Chung J, Lee K 1997 Optimal design of rib structures using the topology optimization technique Proceedings of the Institution of Mechanical Engineers, Part C: Journal of Mechanical Engineering Science 211 (6):425-437
[8] Ding X, Yamazaki K 2004 Stiffener layout design for plate structures by growing and branching tree model (application to vibration-proof design) Structural and Multidisciplinary Optimization 26 (1-2):99-110. doi:10.1007/s00158-003-0309-4
[9] Ding XH, Yamazaki K 2005 Adaptive growth technique of stiffener layout pattern for plate and shell structures to achieve minimum compliance Engineering Optimization 37 (3):259-276. doi:10.1080/03052151512331328231
[10] Svanberg K 1987 The method of moving asymptotes—a new method for structural optimization International journal for numerical methods in engineering 24 (2):359-373
[11] Wall WA, Frenzel MA, Cyron C 2008 Isogeometric structural shape optimization Computer Methods in Applied Mechanics and Engineering 197(33-40):2976-2988. doi:10.1016/j.cma.2008.01.025
[12] Cho S, Ha S-H 2009 Isogeometric shape design optimization: exact geometry and enhanced sensitivity. *Structural and Multidisciplinary Optimization* **38** (1):53-70. doi:10.1007/s00158-008-0266-z
[13] Qian X 2010 Full analytical sensitivities in NURBS based isogeometric shape optimization. *Computer Methods in Applied Mechanics and Engineering* **199** (29-32):2059-2071. doi:10.1016/j.cma.2010.03.005

IOP Conf. Series: Materials Science and Engineering **234** (2017) 012022 doi:10.1088/1757-899X/234/1/012022

Development of Modular Outdoor Furniture Product Using Lightweight Concrete for Public Parks in Surabaya

Grace Mulyono[1], Diana Thamrin[1] and Antoni[2]

[1] Interior Design Department, Petra Christian University, Siwalankerto 121-131, Surabaya, Indonesia
[2] Civil Engineering Department, Petra Christian University, Siwalankerto 121-131, Surabaya, Indonesia

[3] Corresponding author: gracem@petra.ac.id

Abstract. The development of public parks into green city facilities in Surabaya has triggered the need of outdoor furniture designs that can resist the tropical wet and dry weather conditions while also having a certain mobility to support flexible park arrangement. However, present furniture designs made of concrete material are generally heavy and immovable. Flexible designs are needed for various activities that can take place at the same time such as sitting and playing, and to support changes in arrangement to keep the green open spaces attractive from time to time. This research develops the idea of a modular outdoor furniture design using cellular lightweight concrete (CLC) as the main material as a result from observing its resistance towards weather change and its relative light weight. It starts with analysis of problems, formulation of design concept, creation of design alternatives, selection of design, calculation of mouldings, adaptation of design to the mouldings and production of a scaled mock-up using CLC. Findings of this research reveal that the modular design along with the CLC material used not only support the flexibility of change in function and arrangement but also make these furniture resistant to the hot and humid weather of Surabaya.

1. Introduction

The development of public parks into recreational areas in Surabaya has triggered a need for new physical facilities, such as outdoor furniture, to attract people towards the green open spaces. As stated in the *Raperda RTRWP Jawa Timur* (the Indonesian Regional Land Use Regulation Draft of East Java) year 2011-2031, green open spaces need to be maximized as an attempt to achieve the local vision of Surabaya as a 'green city' [1]. Healthy planning of the spaces and its facilities such as furniture and equipment should include a place to improve the health conditions of the elderly, youth, and lower socioeconomic status groups [2]. The designs of the outdoor furniture must be able to resist the wet and dry tropical climate of Surabaya, and can be used for a long period of time. Apart from that, these furniture should be flexible in arrangement, use and management such that they can not only bring variety and interest both architecturally and functionally, but also support the dynamic activities that can take place in public parks, such as sitting, socializing and playing. The growth and development of the city has also changed the meaning of the park in daily life [3]. Hence, in this research, modular multi-functional systems of outdoor furniture, using light weight concrete as the main material, are proposed in order to meet those needs.

The background problem in this research is how to produce designs of modular outdoor furniture that are climate-resistant in the public parks of Surabaya city. Parks and green spaces are of significance at the present time as key elements of individual and social life quality and as supporters of a sustainable city [4]. Through various analysis of outdoor materials, lightweight concrete has been chosen as the main material because of its light weight as well as its resistance towards ultraviolet rays, climate, termites, rust and impact force such that it is relatively safe for both adults and children. Lightweight concrete can be moulded into various shapes and finished with many kinds of coats such that it promotes many possibilities of design in terms of form and visual appearance.

Conventionally, the designs using ordinary concrete as outdoor furniture are usually heavy in weight and thus are mostly attached to the ground or wall, inflexible and cannot be easily moved. By using light weight concrete, the furniture can be arranged and moved by the local park managers according to the needs. The basic material used in this research is cellular lightweight concrete (CLC) that is produced by mixing fine form of air bubbles into concrete mortar [5]. CLC is used because it can be moulded easily, is water resistant, have a relatively high durability, relatively cheap to produce and hence suitable for use in public spaces.

2. Research method

The several steps conducted in developing modular outdoor furniture designs using lightweight concrete are as follows:

- Analysing various conditions of green open spaces and adapting functions and anthropometry to the potential users (children and adults). The design factors of the urban furniture includes colour, harmony with the environment, shape, form, ergonomics and its convenience, has had the highest impact on the citizens' perceptive [6].
- Formulating a solution to address the problems observed in the field through a design concept and applying it in the design alternatives. The module design must be formed together with the formulation of concept so that in the design process, the representation of design would become more distinct [7].
- Selecting the most appropriate design from the design alternatives according to the needs and the production process. The moulding technique, volume and weight were calculated in order to create the flexibility of modular design. From that calculation, the design form was again adapted. A scaled mock-up was produced to form and test the modular system [8].

In the production process, the researchers explored the material mixture, as well as the moulding and finishing techniques that could be used. Through the application of the selected module, feedbacks were drawn in order to evaluate its effectiveness in answering the design objectives and problems determined prior to the design process.

3. Design development

3.1. Analysis of data

Through the observation of the user activities, it was found that the furniture should be able to accommodate the various activities that could take place in public parks such as: sitting while chatting, playing, reading, using electronic devices such as tabs, laptops and mobile phones (Figure 1). Playing facilities were limited to children of age 3-12 years. Meanwhile, sitting facilities were used by a variety of people from children to adults. From the measurements taken from twenty adult users and ten children users, along with the data of anthropometric standards according to Dainoff [9] and Mirer [10], two standard measurements were determined and applied in the design.

From all the data collected, the 5P, or the 5th percentile data- the smallest value, was used to ensure that the design could be used by the whole population. The first standard data was the popliteal measurement of adults and children which are 37.8 cm and 19 cm, respectively. The popliteal height was used as the standard measurement of the sitting facility and was maintained even in the development of the module into other functions besides sitting.

The coating or finishing of the material was also an important factor of consideration in this research. Outdoor coats must be used as the furniture coats observed in the public parks currently have mostly exfoliated, disturbing the appearance and safety of the material. The use of concrete, which is hard in nature, is advantageous because it also cuts down the duration of sitting and thus increases the interchange of users, which is suitable for a public facility. A variety of conditions have been observed in the public parks of Surabaya. Among them, it has been found that some are located in limited areas of land among busy urban streets and thus the position of sitting and playing facilities must consider multifunction designs that could not only make good use of the limited space horizontally but also vertically.

Figure 1. Analysis of user activities and atmosphere of Public Parks in Surabaya.

3.2. Programming and concept

From the analysis, a few design concepts have been formulated in this research as solutions to the problems observed:

3.2.1. Form. Geometrical forms were used to produce modular systems with higher flexibility. Apart from that, geometrical forms can balance the generally organic characteristics of the public parks and can also function as visual focal points through the contrast. In designing urban furniture, principles such as discipline, unity, integration and harmony, balance, rhythm and visual strength must be considered [11].

3.2.2. Function. The modular system was adjusted to the functions of sitting and playing.

3.2.3. Dimension. The popliteal dimensions of both adults and children were applied as the standard dimensions of the modular design.

3.2.4. Material. Major principle which have been observed in the design and deployment of urban furniture are to respond to the needs of the environment [12]. Observing all the flaws of the materials that have been used previously in public parks, CLC was chosen as the main material in the design due to its resistance to weather changes and manageable light weight that can support the flexibility of arrangement. The use of this apparently durable material can prolong the use of the furniture and thus reduce the need of repair and maintenance from time to time as have been observed in the current conditions of the furniture in public parks.

4. Modular design for outdoor furniture

4.1. Design Alternatives

From the concept formulated and determined, three design alternatives were produced (Figure 2). The design alternatives were then criticized and analysed in terms of their strengths and weaknesses. From the analysis, the third design alternative was chosen because of the flexibility it could offer in terms of its arrangement into both sitting and playing facilities. The module is also simple in form which is considered advantageous as it can be arranged into a complex form of modular system. From the perspective of design, in terms of form and dimension, this module has the most positive potentials. However, when analysed in terms of production technique and material, this form is not suitable for light weight concrete. The strength obtained from a solid mass form was deemed more favourable as it hinders the need of using frames for construction and stability. Thus, though the third design alternative was chosen, it was further developed into a more bulky solid form in order to strengthen the light weight material used.

No	*Design Alternatives*	Analysis and Discussion	Conclusion
1		• The modular system in this design can only be arranged horizontally and hence require larger areas of space. • The organic form inhibits possibilities for more variation in terms of modular composition. • Each design form can only be used for one function.	DESIGN DEVELOPMENT
2		• The design form is more geometric and the modular arrangement can save up more space compared to the first alternative. • Vertical arrangement is possible. • The modular composition can accommodate multiple functions for both sitting and playing.	DESIGN DEVELOPMENT
3		• The design is more geometric, the modular arrangement can save up more space compared to the first alternative. • Vertical arrangement is possible. • The modular composition can accommodate multiple functions for both sitting and playing. • The geometric form is much simpler than the second alternative and hence exhibits possibilities for more variety in modular composition.	**SELECTED DESIGN**

Figure 2. Design alternatives produced and the selected design.

4.2. Selected Design

The module with a squared form and openings was chosen and further developed into a more solid and bulky form (Figure 3). The geometrical form was maintained in order to promote more variation in the modular arrangement system, while the dimensions adopted the two varying popliteal measurements of adults and children. The higher part of the form could be used for adult sitting while the lower part could be used by children for sitting as well as function as a step for climbing. With the solid module, the design was then developed into various modular arrangements through exploration of forms using scaled models. In the final design, two dimensions were obtained based on the standard of popliteal heights of both adults and children (Figure 4).

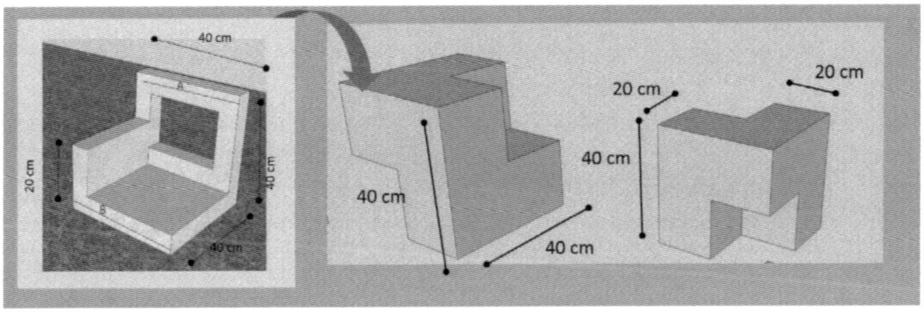

Figure 3. Design development process.

The application of the standard dimensions based on the popliteal dimensions of adults and children, 37.8 cm and 19 cm respectively, resulted in a formulated dimension of 40 × 40 cm in the design. The use of equal dimensions on both sides of the module ease the combination of the modular system. With this dimension, each module weighs about 33.6-38.4 kg. This weight is sufficient and suitable for arrangement change and lifting, and thus is good for mobility, easing the public park managers.

The development of the selected design was then continued with the making of a study mock-up of 1:10 scale using plasticine (Figure 5). The first prototype was made in order to find modular compositions based on form and function. From the various modifications done, several modular systems of sitting and playing facilities were discovered. This prototype provides a limited overview of the effectivity of dimensions and compositions of modular system that can be achieved.

The next step involved the making of a second prototype with light weight concrete with the scale 1:4. Based on the modular compositions obtained in the experimentation done on the first prototype, the second prototype was tested using trial and error in terms of form and material.

Figure 4. Various applications of the selected module.

Figure 5. Modular compositions made tested with plasticine.

4.3. Design Evaluation

The design evaluation involved the feasibility study of the product in terms of the modular systems formed, stability and the strength of the design. In this process, it was found that the cavities or openings designed on the upper and lower part of the module ease the interlocking system of the module. However, they cause the outer modules to be unlocked which can trigger the modular composition to collapse. Pressure from one side of the module may cause the unlocked modules to topple down. From the evaluation conducted, at least two modules should be used, in which one module has two cavities while the other module functions as a locking module having only one cavity (Figure 6).

From the prototype made, several modular systems were then produced that could function as sitting as well as playing facilities. The system was then evaluated in terms of form, joints and suitability of function. As shown in Figure 7, composition no.1-3 dominantly function well as sitting facilities, both for adults and children. Whereas, composition no. 4-5 can function as sitting as well as playing facilities. With the second module (Figure 6) functioning as the locking module, the modular system would be stable and can also be arranged vertically.

Figure 6. The two basic modules that can be applied in the selected design

Figure 7. Development of modular compositions from the second prototype.

4.4. Material Exploration

Exploration of material mixture were conducted to yield a concrete mixture that is light-weight and not brittle. Cellular Lightweight Concrete (CLC) is a light-weight concrete material that uses foam generator to produce fine air bubbles to be mixed into concrete mortar having advantageous benefits such as high workability, lightweight, use of less material, low water absorption and a good resistance to change in temperature. The air bubble in CLC can be as high as 60-70% by volume and this results in a lightweight material, consequently decreasing its compressive strength [13]. However, consistency in the light weight of the CLC brick can be maintained by adjusting the weight and volume of the foam.

In order to make the CLC, composition of mixture uses about 300 kg/m^3 of cement, 300 kg/m^3 of fine sand, added with water and foam to yield a wet density of about 800 kg/m^3. From the feasibility test of the material, the compressive strength acquired was about 2.5-3 MPa with a water absorption value of 25% of the weight. The more foam used in the mixture, the less the weight of the resulting

CLC. However, if the weight is too light, the concrete would be unusable because it would be brittle and have low strength. This is proven by fact that a light brick with a weight of 500 kg/m^3, when removed from the formwork, would result in very porous material and can no longer be used. Hence, the ideal weight for lightweight concrete in this design is between 700-800 kg/m^3.

4.5. Production Technique and Process

Through the evaluation of the material composition conducted, a cast system with wooden boards have been determined as a suitable system for making the mouldings. Wooden boards were chosen because of their low cost compared to iron and silicon mould. The wooded boards must be coated with wax so that they can be detached from the mouldings and used again. For mass production, certainly it will be feasible using steel mould for faster casting cycle and longer mould life time.

The concrete that has dried must be polished to smoothen the surface and to minimize sharp edges so that they could be safely used by adults and children. This is then followed by the finishing process by sealing and polishing the small air bubble that was visible in the surface using cement. Grey colour, the basic colour of the concrete was chosen. The outer surface should then be coated with weather-resistant coat having a doff texture as material protection. The various shades of grey can produce attractive natural tones of composition that complement with the natural settings of the green spaces.

5. Concluding remark

In designing urban outdoor furniture, the activity of its users as well as the durability of the material need to be considered. Designing the furniture as a modular element, improve its usability in adapting with the change of user activity. Lightweight yet sturdy material of CLC was selected as the main material, as it has low economical value which also prevents the loss of properties.

The use of cellular lightweight concrete, which is hard in nature, is advantageous because it cuts down the duration of sitting and thus increases the interchange of users, which is suitable for a public facility. A variety of conditions have been observed in the public parks of Surabaya. Among them, it has been found that some are located in limited areas of land among busy urban streets and thus the position of sitting and playing facilities must consider multifunction designs that could not only make good use of the limited space horizontally but also vertically.

Acknowledgement

The authors acknowledge the financial support of Applied Product Research Grant from Ministry of Research, Technology and Higher Education, Indonesia, in this research study.

References

[1] W Widigdo and I K Canadarma 2010 Surabaya as garden city or green city *Nasional Seminar on City Architecture "Life and Living in Surabaya"* Surabaya, 27 Mei 2010
[2] J Maas, R A Verheij, P P Groenewegen, S deVries and P Spreeuwenberg 2006 *J. Epidemiol Community Health* **60** 587–592.
[3] G Atanur 2015 *ArchNet-IJAR* **9** 1 247-260
[4] C W Thompson 2002 *Landscape and Urban Planning* **60** 59–72.
[5] Antoni, R Jos and M M Lukito 2011 *Proc. 4th ASEAN Civil Engineering Conference.* Yogyakarta
[6] R Kargar and M Ghassemi 2016 *Journal of Civil Engineering and Urbanism* **6**(2) 40-47
[7] K N Otto and K L Wood 2003 *Product Design* (Pearson Education Asia Limited PRC)
[8] CC Huang 2000 *Proc. Natl. Sci. Counc. Roc (A)* **24** 3 149-165
[9] M.J. Dainoff 2008 *Ergonomics and Psychology Developments in Theory and practice* (Boca: CRC Press) 12-17
[10] S B Mirrer 1987 *Children's Environments Quarterly* **4** 3, http://www.jstor.org/stable/41514639
[11] S Zeinali and B B Babaei 2015 *Int. Research Journal of Management Sciences* **3** 1 9-13
[12] R Afsari, S R Mousavi, Z Aghayari, M Shiripour and V Bakhshi 2014 *Journal of Civil Engineering and Urbanism* **4** 109-113
[13] D K Panesar 2013 *Construction and Building Materials* **44** 575–584.

IOP Conf. Series: Materials Science and Engineering **234** (2017) 012023 doi:10.1088/1757-899X/234/1/012023

Design of Backpack to Aid Elderly for the Mazu Touring Procession in Taiwan

F L CHAO [1], Y C HUANG[1], J Y SU[1], C L SUN[2] and C C CHEN[1]

[1] Department of Industrial Design, Chaoyang University of Technology, Taichung, 436, Taiwan, R.O.C.
[2] Department of Social Work, Chaoyang University of Technology, Taichung, 436, Taiwan, R.O.C.

flin@cyut.edu.tw

Abstract. The Dajia Mazu Touring Procession is a 9-day long religious event held annually. However, for the elderly participants, it is a big burden especially in regards to physical strength. The goal of designing backpack is to reduce the physiological stress of elderly during the procession. Firstly, physical parameters were measured to explore the dimension parameters by testing. The height of the chair is different from that of the kneeling pad; a smooth curve was chosen to coordinate the two as the main outline of the backpack. Secondly, material selections based on following limits were considered: (1) acceptable weight and size, (2) intermediate price and (3) a design that is fitting to the Dajia event. The material and structural strength were evaluated for wood, bamboo, stainless steel. Two design concept were proposed, wood is selected for construction and testing by users. The texture of the backpack is Rush grass, it was built successfully to cover the backpack's external surface to meet local culture features.

1. Introduction

The annual eight-day Mazu Holy Pilgrimage, which usually takes place in April, is a religious festival that begins at Zhenlan Temple (zhènlángōng) in Dajia (dàjiǎ), Taichung County to Fengtian Temple (fèngtiāngōng) in Xingang (xīngǎng), Chiayi County [1]. It perhaps is one of the most bizarre of all Taiwan's religious festivals. The pilgrimage traces its roots to the early nineteenth century, when Taiwanese pilgrims would cross the Taiwan Strait to the Mazu "mother temple" in Meizhou in Fujian every twelve years. The core procession comprises a series of palanquins that ferry Mazu and other senior Taoist deities 200 kilometers through rice fields and small villages. Stops are made at smaller "branch" temples to enhance the power of local deities. The statue remains in Xingang for a day of celebrations for Mazu's birthday before embarking on its four-day journey back to Dajia.

Pilgrims usually participate in the full trip by walking. For the elderly pilgrims, the burden of physical strength is especially consequential. Relying only on perseverance in the long journey, the elderly occasionally need to kneel in prayer for blessing of their families. To comply with formalities and tradition, one will have to kneel with the head touching the ground, which is an especially difficult task for the elderly.

Supporting of the core muscles group is required in order to go from kneeling to standing position. The abdominal muscles need a strong contraction of the organs within the abdomen to produce some pressure to help evenly distribute the force to the entire spine. Usually people train muscle endurance and core muscle groups under stable circumstances. If the abdominal muscles do not have enough contraction force, muscles on the lower back may easily become fatigued or sore.

By kneeling, the knee join may also be damaged. Sports knee-pads are commonly used by football goalkeepers, volleyball players, as well as male and female hip-hop dancers. Thickened sponge knee pads are fabricated with highly elastic knitted fabric (which will not shift in position during the bending), with sponge lining added to provide protection for the knee. One of the shortcomings of sponge filling knee pads is that it causes pressure around join during walking.

The kneeling stool can be used in temples, and usually consists of soft material wrapped with leather. High-density foam is often used as the soft material to reduce the impact as the knee touches the pad. Through the sloped surface and the foot support bar, the stool height is increased to 17 cm to reduce the joint pressure (figure 1). Apple News has reported [2] the launch of a newly designed "cool bench" for the Dajia Mazu Touring Procession, which allows the believers/pilgrims to sit down and rest at any time (figure 2). The bench is designed to be similar to a drum that can be tied to the waist. It is light weight and has an elastic surface for one to sit on. Because the stool is usually tied to the waist, and pilgrims walk most of the time during the procession, pilgrims will have to walk with the attached stool shaking behind their buttocks, which may not be aesthetically pleasing.

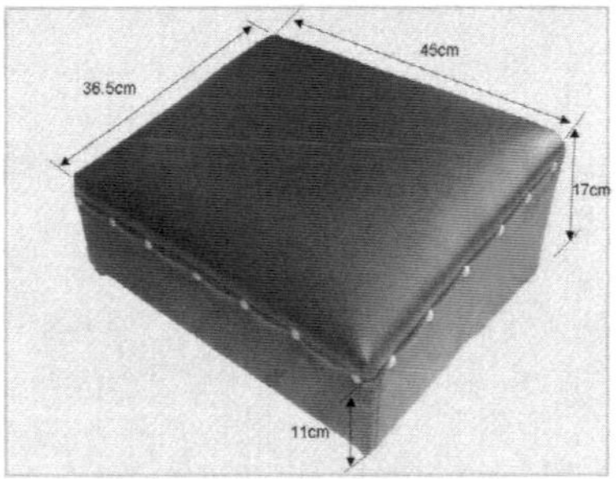

Figure 1. Kneeling stool.

Figure 2. Cool bench [2] for the touring procession.

2. Design parameters estimation

2.1. Previous study

Many scholars have studied the design of the seat to reduce the body load while getting up. In designing household chairs for the elderly to sit, the scholars discovered that (1) the elderly often have to rely on other objects to assist them in standing up, (2) low seat surface will make it more difficult for elders to stand up, (3) sometimes elderly people fall onto chair during standing. The study developed solution by means of a pneumatic bar mechanism to assist the elderly in standing up, which was found to require less effort for the elderly [3]. In the study of the dynamic needs for the elderly of the seat; the researchers found that the key parameters are the seat height and armrest height. They also study the chair armrest design, and found out that as the age of an elderly increase, the degradation of muscle strength makes it difficult for them to sit and move [4]. The best armrest position and height ranges are between 65 to 70 centimetres.

The items that a Pilgrim will need during Mazu touring procession include simple carry-on items such as cutlery, toiletries, clothing, medicine, water and food, most likely packed in a backpack. Since people will carry these things as they walk, the backpack weight should be light. When the item are placed in the backpack, the backpack can be converted into a temporary chair or kneeling pad. The size of the chair and kneeling pad needs to be optimized both in terms of convenience and supportiveness. To avoid interfering with walking, handrails should best be able to be collapsed. The location where the backpack and item can be opened and closed should be chosen carefully to provide convenience of access.

2.2. Chair and kneeling pad experiment

The experiments were planned according to the individual needs of the seat and the size of the kneeling pad. If the subject's height is 160 cm, with sitting height of 28 cm, kneeling height of 13 cm, handrail height of 14 cm, then the total relative height is 42 cm [5].

The height of the chair and the height of the kneeling pad are different. Thus we take both height and use smooth curves to harmonize between the two parameters, to use as the main contour of the backpack. By sitting, one has his or her buttocks contact the chair surface; however, in kneeling, one has his or her knee in point contact with the knee pad, and will have to use the feet as a fulcrum to remain in balance. Thus the kneeling pad is designed to be lower than the chair. The height of the seat is also limited by the environmental constraints that occur as the user is walking. The handle is set at the levelled area at the top of the seat. The handler height is changed at the chosen position in order to determine suitable handle height. The height of the handle bar, relative to ground level, was changed to the thickness. The handle bar height relative to the ground level is 42 centimetres, and that of the handle bars height relative to the top of the chair is 13 centimeters [5].

From the experiment, it was discovered that, for subjects from different age groups, the change in joint angle over time from kneeling to standing is clearly different. The response change corresponds to elderly people is slow and there are slight shakings present (mild slope). This indicates either a lack of muscular strength or loss of flexibility in the ligaments, thus leading to a bit of struggle in standing up [6].

3. Concept design procedure

The proposed backpack auxiliary object can be used to carry items, and can be used to sit. The parameters that were researched can be applied on the design of the auxiliary object; the backpack needs to be made with lightweight materials, and auxiliary parts should be added so that the elderly can have support while squatting down or standing up, reducing the load on the knee. The combination of the seat, the objects for support, and the backpack makes the item convenient to use for the user. The curved appearance and the height difference of the item allows kneeling for prayer, and sitting for rest. Handle bars are added on the two sides of the item, so that the elderly have support while kneeling down or standing up. Thus, the physiological load can be reduced in the procession.

Material selections based on following limits were considered: (1) acceptable weight and size, (2) intermediate price. The material processing and structure strength were analyzed for wood, bamboo, stainless steel and plastic (table 1). Wood strength is not high, easy processing, the use of reinforcing ribs can maintain sufficient strength, but will also increase the weight. Curved wood technology can maintain sufficient tension; it requires tooling template and molding tool in manufacturing.

Bamboo is inherently lightweight and resilient, and can withstand pressure or force in a particular direction, but the strength is rapidly reduced when the joint is poorly handled. It also requires special techniques for bending and using tension to provide support. Plastic, for example Acrylic, can form continuous curved surface structure under specific temperature and pressure conditions and providing sufficient strength. The forming cost is high; costs can not be shared in a small amount of production. The use of lightweight steel pipe structures with canvas tension property can provide durable support. The initial processing costs can also be controlled in a reasonable range. Owing to the initial prototyping cost, the wood and stainless steel were selected during brainstorming process.

Table 1. Material selection considerations.

Material	Weight	Strength	Processing	Cost
Wood	+	+	+	++
Bamboo	++	++	-	-
Plastic	+	+	-	-
Steel	++	++	+	++

4. Design A: stainless steel

Backpack topical material functions require waterproof, anti-wear. It can be divided into two categories, nylon or man-made fibres, each with different characteristics. Suggested fibres are more than 1000 denier for durability. The benefits of canvas are contact and comfortable. For waterproof consideration, it can be added another inner surface with waterproof protective material. Although the canvas is heavier but more suitable materials, it can fulfil the need of outdoor activities and durable in pressure.

Considering the balance of the spine, the left side of the spine will bend to the left, and ultimately will oppress to the lumbar spine. The shoulder strap of backpack is easy to fall, the user often unconsciously raise the shoulder to stabilize the shoulder strap. So that the shoulder muscles stay in a state of contraction, which lead to shoulder pain and joint muscle strain and causes shoulder and back problems. Carrying heavy items, it is appropriate to use double shoulder strap. By utilize good back plate (figure 3), the body can in line with the body's three-dimensional structure, to strengthen the protection of the lower back and shock absorption. The guide grooves with air convection channel that made by heat conducted material also can improve heat dissipation ability.

The proposed concept design has dual shoulder straps (figure 4). Through the reinforced material within the backplane, it can support sitting or kneeling posture. The 2D computer product sketch is presented in Figure 5, kneeling posture schematic diagram is also shown. In the future, through part modularize it can provide required sitting and kneeling support.

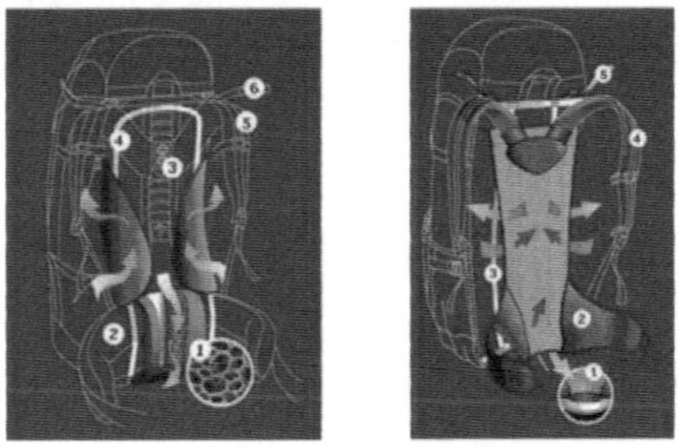

Figure 3. Guide grooves with air convection channel.

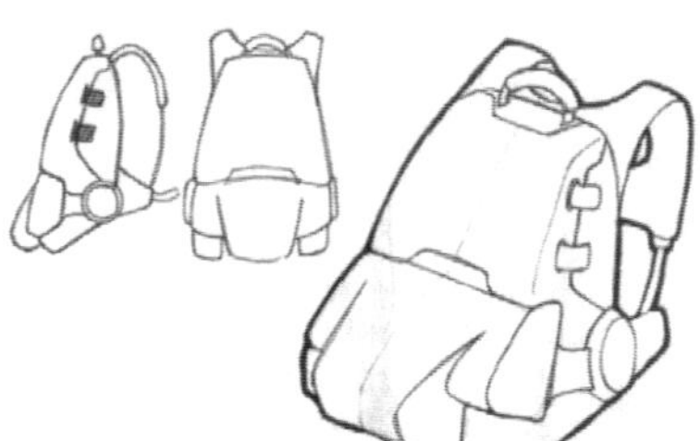

Figure 4. Design A: inner structure is stainless steel, outer surface is canvas.

IOP Conf. Series: Materials Science and Engineering **234** (2017) 012023 doi:10.1088/1757-899X/234/1/012023

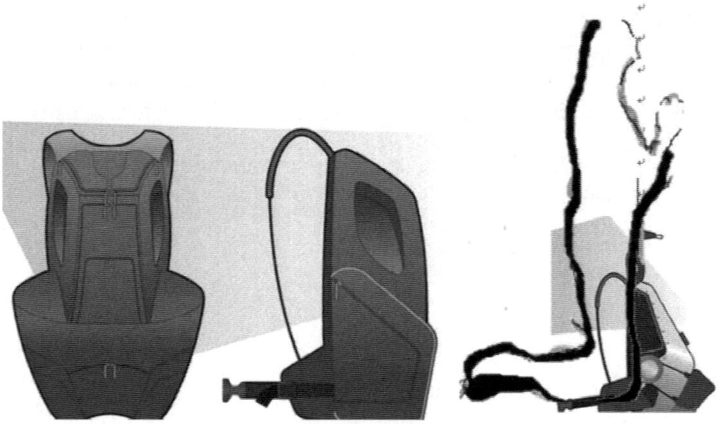

Figure 5. Kneeling posture schematic.

5. Design B: wood

5.1. Support handle

As the backpack is placed on the ground, by changing the direction the backpack may become a seat or a kneeling pad. On the two sides of the backpack there are handles that can assist in standing up. As the user wants to stand up, the auxiliary handle can be extended outwards (figure 6), and after it is stabilized it can be held as a handle to assist the elderly in standing up. In order to avoid interfering others during the march, the handle have features that allow folding, and can be opened during use (figure 7).

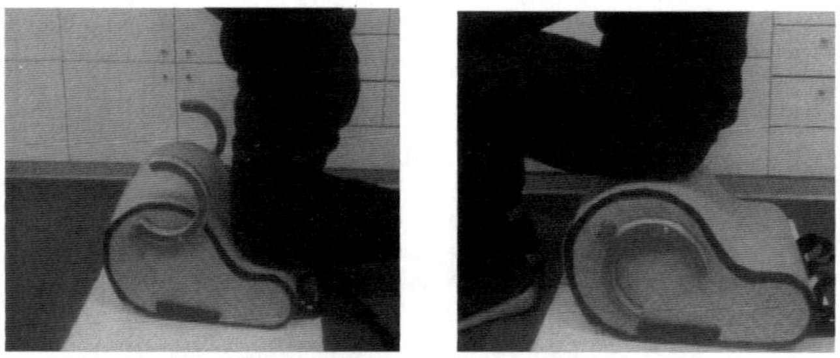

Figure 6. Handle usage and the kneeling pad.

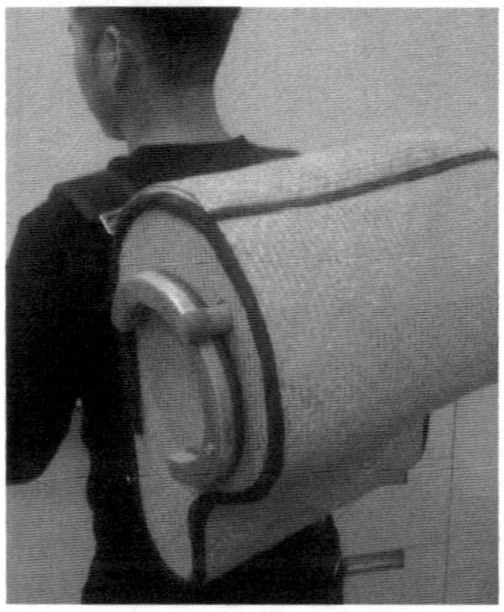

Figure 7. Combination of the seat, support, and the backpack

5.2. Surface texture of Rush grass

To incorporate spatial capacity, appearance in sitting and kneeling, and to assist in standing, two design concepts were selected for model construction and user testing. The texture of the backpack is Rush grass (*Cyperus malaccensis*) [7]. Historically, local people use wild ridge grass with sub-root transplanted to paddy fields, and after harvested treatment, transform grass into thin strips, and wove them into beautiful straw mattresses. Triangular cross sections of grass were utilized in hand-woven weaving, producing a soft texture that exudes a touch of fragrance and cool feeling in summer [8]. In the final design process, rush grass is successfully used to cover the backpack's external surface (see figure 8) to express local cultural features.

Figure 8. Rush grass surface texture by hand-woven weaving (Mrs.Y J Sun).

6. Conclusions

The backpack combines a seat and a kneeling pad, with auxiliary handles to assist in standing up, allowing the elderly to have support while kneeling down or rising up, thus reducing the physiological load for elderly during the Mazu Touring Procession. The curved appearance and use of lightweight materials does not increase the weight by too much; the item will have affordable weight, size, and price, and is designed with an appearance that is compatible with the local image, and thus can be accepted by consumers, which is the goal that the designer pursuing.

References

[1] Http://travel.cnn.com/taiwan-mazu-religious-festival-pilgrimage-226351/
[2] Apple Daily April 2016 http://www.appledaily.com.tw/realtimenews/article/new/20160408/834412/.
[3] Doorenbosch C A, Harlaar J, Roebroeck M E and Lankhorst G J 1994 Two strategies of transferring from sit-to-stand: the activation of monoarticular and biarticular muscles *Journal of Biomechanics* **27** 1299-1307
[4] Dubost V, Beauchet O, Manckoundia P, Herrmann F and Mourey F 2005 Decreased trunk angular displacement during sitting down; an early feature of ageing *Physical Therapy* **85** 804-81
[5] Chao F L 2017 Parameter estimation of assisting objects to aid elderly for Mazu touring procession *Ergonomics Society of Taiwan conference* Quemoy R.O.C 100-105
[6] Khemlan M M, Carr J H and Crosbie W J 1999 Muscle Synergies and joint linkages in sit-to-stand under two initial foot positions *Clinical Biomechanics* **14** 236-246
[7] Http://tropical.theferns.info/viewtropical.php?id=Cyperus+malaccensis
[8] Http://www.mdnkids.com/taiwan_library/index2.asp

IOP Conf. Series: Materials Science and Engineering **234** (2017) 012024 doi:10.1088/1757-899X/234/1/012024

Elderly's Family Life Supplies - Innovative Chinese Checkers Game Board

Fanglin CHAO [1]

[1] Department of Industrial Design, Chaoyang University of Technology, Taichung, 436, Taiwan, R.O.C.

flin@cyut.edu.tw

Abstract: The product design course for industrial design students was implemented in our university which spans 9 weeks. Throughout the creativity rules and field study, students achieve high standard on problem identification and concept generation. The prototype test with elderly in design projects is helpful to make students with deeper understand user demand, which in turn enhance the concept further. Traditional Chinese checkers are redesigned using special checkers with different height or shape and specific rules to increase user interest and game diversity. Game is more challenging due to location weighting on score calculation to planning its strategies. Redesign Chinese checkers game board include reconfigurable board and several shape checkers. Checkers has 3 parts: standing ring body, the base body, both support the side holding structure. The body shows slightly concave to facilitate the fingers hold. The upper portion of the body is provided with different shapes extension section which can be engaged with base body. Player move the checker to the opposite target area. When one of player moved all the checkers to the opposite target area; they shift to the scoring calculation stage. The participant may develop specific strategy to gain higher score by maximized weighted checkers into its target block regions.

1. Introduction

1.1. Creative design

Design is often a highly multidisciplinary subject. There are many stages in the concept generation to consider user requirement and technologies feasibility [1]. It will be involved in both concept generation and the evaluation phase. The education institutes usually do not have enough experts to enable a true multidisciplinary guidance. Observation and field study are required to enhance feasibility of design thinking.

Empirical data on design processes were obtained from a set of protocol studies of nine experienced industrial designers [2]. Aspects of creativity in design are related to the formulation of the design problem and to the concept of originality. They also propose refinements to the co-evolution model, and suggest relevant new concepts of 'default' and 'surprise' problem/solution spaces. In [3] paper uses three case studies to discuss the importance of a total integration between marketing research and design at the concept stages of new consumer product development and throughout the design development and marketing process of the product. In [4], the designs for behaviours change by testing the potential of priming and collective concerns were reviewed.

Incorporating into subconscious priming stimuli for each of the major senses; the studies showed a significant effect across all senses.

1.2. Elderly
Ageing populations are on the rise worldwide. The 21st century is destined to have an increasingly ageing population of elderly individuals. Health is one of the most determinant elements in quality of life in the elderly. Therefore, strategies are needed in order to improve their quality of life [5]. Retired elderly individuals have more time for leisure activities and may devote themselves to their preferred hobbies and activities. Being occupied may prevent the elderly from falling into a sedentary lifestyle and to slow down the effects of ageing. A moderate correlation was found between being active, be it with daily tasks or leisure-related activities, and a higher degree of perceived health [6].

Currently there is around 0.25 million Taiwanese with dementia, and in 2030 this number will double to around 0.47 million. This not only affects the patients themselves, but also their families as well. By utilizing toys, it is hoped that the patients can engage in activities and slow down the degradations caused by dementia.

1.3. Chinese checkers
Chinese checkers game boards are often played by the elderly. Due to its simple rule, after players grasp its knack, it's easy to win and thus they would lose interest. Checker pieces are often in conical shape, which is difficult to hold. Plastic pieces can be unstable, making them easy to tip over by accident, thus affecting the game.

The shape of the existing checker piece is conical, which is not easy for the finger to hold, elderly may accidentally push down or touch other pieces. In addition, some of the elderly have cataract problems. For cataract patients, purple, green and red colored lights can be more easily seen, while blue light is less visible. This is due to the light being filtered out in the eye, therefore eye diseases such as macular degeneration or other degradations must be considered.

2. Design procedure
The product design course for industrial design students was implemented in our university which spans 9 weeks. Throughout the creativity rules and field study, students achieve high standard on problem identification and concept generation. The course is provided in Department of Industrial Design as the first project for young designer (year 2 students). Students are required to design elderly utility that without high technology elements. The product concept of checker is provided here to indicate brainstorming and creative thinking process.

Basic creativity rules (such as Reverse, Transfer and Combine) [7] and field trip was introduced in the first 3 weeks. The mind set help extend thinking domain were introduced with visual representation [8]. Then each student choose individual problem to design through both conceptual drawing and hands-on prototyping (during 4 weeks). Students created their own design, teacher guide them how to judge and balance each design factors. Finally, they take two week in presentation and panel design.

3. Concept design

3.1. Innovative checkers
Innovative checkers have different height or forms; they follow different scoring rules which can increase the challenge. The innovative Chinese checkers has a new scoring system related to piece's tallness or type to make it more fun to play. Pieces of different tallness or type have to be moved to the designated area for more points. The colors of red, green and yellow are chosen for the color of the pieces.

Tallness of the pieces represents family seniority. The taller senior pieces are places at the end of the section. High score is achieved not only by moving pieces to the opposite target region, but also

IOP Conf. Series: Materials Science and Engineering **234** (2017) 012024 doi:10.1088/1757-899X/234/1/012024

have the senior pieces moved to the end of that region (figure 1). Players now have to plan their strategies to move senior pieces out first, and use shorter junior pieces as the jumping platforms.

The shape of the pieces is redesigned for elderly so it is easier to hold, more stable and less likely to tip over others during action. For the shape to be easier to be held, the curvature of the pieces is designed to be greater, so fingers can more easily grasp. When moving the pieces, it is more stable (figure 2), so it is less likely to drop and affect other pieces.

The concept generation process refers to the following creative rules:

- Reverse: checker is usually same size and shape, concept to use asymmetric or non-uniform shape to increase the challenge.
- Transfer: select the traditional family "hierarchical structure" as the main element, transfers the element into checker's pieces.
- Combination: checker frame is modified and combined with scoring calculation rules to increase the challenge.

According to the co-evolution model [2], the creativity rules support in extended thinking domain which suggest relevant new concepts of 'surprise' problem/solution spaces. In [4], incorporating into subconscious priming stimuli for each of the major senses; the studies showed a significant effect across all senses. The checker design provide more sense involved such as: sight (the shape represented family member), touch (easy for grasp), and thinking (strategy).

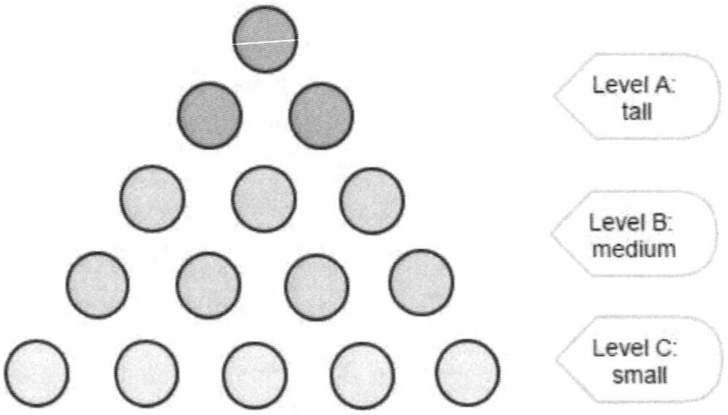

Figure 1. Chinese checkers placement original concept.

Figure 2. Shape of the pieces needs redesign.

3.2. Play method and scoring

On the aspect of play approach, rules are designed to increase the fun factor and ways of thinking, by using pieces of different height or type. In figure 3, in order to move the senior pieces to the end of the opposite region, players must plan strategies accordingly. When one player moves all pieces to the opposite region, the game ends. Players stop moving any more pieces and enter the stage to calculate the final score.

During the calculation step, besides counting all pieces entering the target region as one point, there will be weighted points too. Those senior shaped pieces entering the inner section of the target region can gain more points. Even though one player has already moved all pieces to the opposite region, it does not necessarily guarantees victory. Once the weighted scores are considered, the final score might favour the other player.

For example, the seniority of two levels, depending on setting by both sides there are 3 or 6 higher level pieces, this setting is simpler and more suitable for beginners. After the players are familiar with the rules, then they can play the game with three seniority levels. Even though the rules are the same, the player has to manipulate three types of pieces and weighting scores, making the game more complex and suitable for advanced players.

Play method 1:

(1) Pieces need to go back to the same color / same block region,

(2) Block regions include (large → medium → small) arrangement.

Play method 2:

(1) Pieces correspond to the same pattern, block region (large → medium → small) arrangement,

(2) Given the weighted score, the player with highest accumulated score wins, where A region gains 5 points, B region 3 points, C region 1 point.

Unlike tradition Chinese checkers, players decides on the division with weight, and with this plan move the pieces strategically, which also is more challenging. Using figure 4 as the example, the scoring of setting with 3 or 6 higher level pieces is as following.

The upper left part of figure 4 shows three senior pieces at the senior section, so the score is $3\times3+12\times1=21$. Lower left part of figure shows the score of $3\times2+13\times1=19$. Consequently, the upper left one is the winner. Upper right part of figure shows the senior group with 6 pieces, and the score is $3\times6+9\times1=27$. Lower left part of figure shows the score of $3\times4+11\times1=23$. Consequently, the upper left one is the winner. Therefore the first to move all the pieces to the target region is not necessarily the winner.

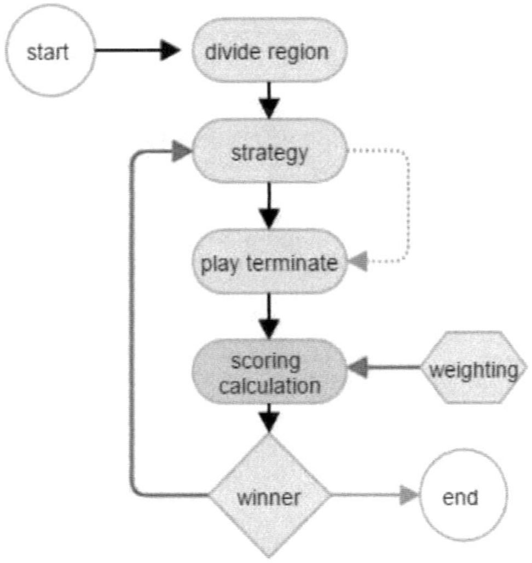

Figure 3. Flow chart: players have to plan their strategies based on weighting.

IOP Conf. Series: Materials Science and Engineering **234** (2017) 012024 doi:10.1088/1757-899X/234/1/012024

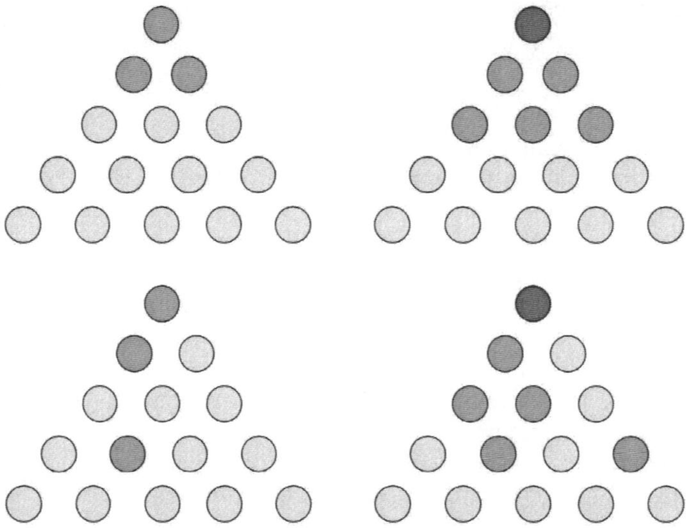

Figure 4. Weighted scores calculation, the upper left (n=3): 3×3+12×1=21, the lower left: 3×2+13×1=19. The upper right (n=6): 3×6+9×1=27, the lower right: 3×4+11×1=23.

4. Structure adaptation

Figure 5 shows structure design of the variable height pieces. The piece includes: one standing ring shaped main part, one basement with mass. The two are locked together at the basement's side. The main part is slightly curved inwards, so it is easier to hold by the finger. The top of main body has one section with a removable extension part, which can inserted into the main body, and locked to the basement to form a piece with variable height. Regarding the design variation of shape, user can select shape depending on geometric form or schematic outline of higher level pieces. One can have a senior shaped or change to have a junior shaped piece, so they can be distinguished from other types.

Secondly, the design of the board must also be adjustable in order to separate relative position of the seniority groups. At the start when pieces are placed, target region has to be marked for different seniority groups. One way to do this is to add a transparent layer on top of the existing magnetic Chinese checkers board. Seniority groups use different patterns of the main colour to represent blocks region. Different groups can be distinguished by different stripe patterns (figure 6), for example the number of stripes; it is easy to distinguish three seniority sections in the target region. When the seniority group region is fixed, the transparent layer can also be fixed on top of the board to avoid instability.

Figure 5. Structure designs of the variable height pieces (by Z. H. Lau).

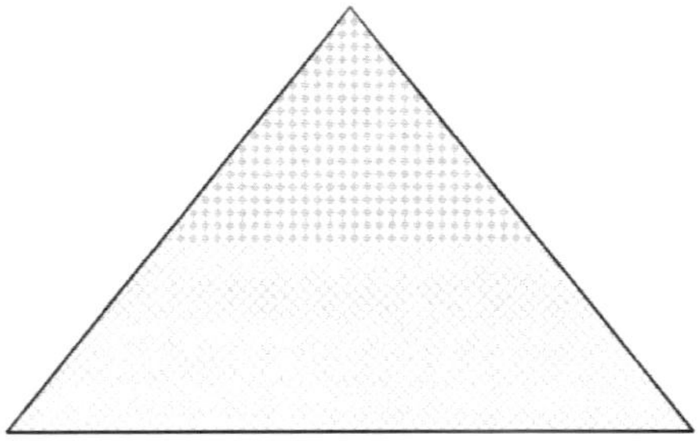

Figure 6. Different region can be distinguished by patterns or shading.

5. Conclusions

The product design course for industrial design students was implemented. Throughout the creativity rules and field study, students achieve high standard on problem identification and concept generation. Toys can enhance interaction and brain activities of dementia patients and slow down the mental function degradation. The elderly likes to play Chinese checkers, which suitable for two or more players. Elderly also likes to be accompanied or to play with grandchildren, so they can chat while playing.

By adding rules about pieces of different height or type, it is hoped that it can increase the fun factor and encourage a more diverse strategic thinking. The change of pieces' shape allows elderly to hold with ease, avoid pieces from tipping over, or touch other pieces by accident during action. The seniority level concept originates from the elderly, but it does not limit the player to only elderly people. The board game can attract players of different age group, and offer good experience.

Acknowledgment

The authors wish to thank Pai Jung Lee in concept discussion, Lecture Z.H. Lau for 3D drawing and Professor T.W. Hu for providing constructive suggestions of the paper.

References

[1] Kumar V 2012 *101 Design Methods: A Structured Approach for Driving Innovation in Your Organization* (New York: Wiley)
[2] Alexander M 1985 Creative marketing and innovative consumer product design—some case studies *Design Studies* **6** 41-50
[3] Dorst K and Cross N 2001 Creativity in the design process: co-evolution of problem–solution *Design Studies* **22** 425–437
[4] Cash P and Holm-Hansen C 2017 Uniting individual and collective concerns through design: Priming across the senses *Design Studies* **49** 32–65
[5] Wilcock A 2008 *Professional Identity. Revista Terapia Ocupacional Galicia* 35-47
[6] Ball V, Corr S, Knight J and Lowis M J 2007 An investigation into the leisure occupations of older adults *The British Journal of Occupational Therapy* **70(9)** 393- 400
[7] Lee M H 1996 *Becoming Creative* (Taipei: Scholars publishing)
[8] Faste R 1995 A visual essay on invention and innovation *Journal of Design Managemen* 1-21

IOP Conf. Series: Materials Science and Engineering **234** (2017) 012025 doi:10.1088/1757-899X/234/1/012025

Towards Enhanced Affective Design: Rethinking the Notion of Design

SuKyoung Kim & Youngil Cho

Sapporo City University, Geijutsu-no-mori 1 Minami-ku Sapporo 005-0864 JAPAN

s.kim@scu.ac.jp

Abstract. Design disciplines have been contributing to shaping the life of human beings, as well as fostering culture and heritage. Design disciplines and research have been rapidly transforming, and not only objects but also services are target of design. This paper reviews design disciplines towards enhanced affective design, which attributes to intuitive knowledge. It aims at rethinking the notion of design to propose a conceptual framework for integrating user experience into objects that strengthen the form and function based design with pleasing.

1. Introduction

"Design is a creative activity whose aim is to establish the multi-faceted qualities of objects, processes, services and their systems in whole life cycles. Design is the central factor of innovative humanization of technologies and the crucial factor of cultural and economic exchange. Design seeks to discover and assess structural, organizational, functional, expressive, and economic relationships [1]". This definition of design makes the point clear that what is design for, and current tendency of design shows what the definition represents. Design contains various social context, influences innovative humanization of technologies, and change economic with new design notions. In the first industrial revolution, new devices and machines started to replace human labour in many industries, so called "mechanization." Water power and steam power replaced human power. At that time, design was for making "new power" which never been existed before but experienced with similar characteristics—for example, horses, which is to ride on, carrying with, moving to, were substituted by the automobile. In the second industrial revolution, also known as the technological revolution, machines were designed more efficiently, so called mass production era started. With the introduction of new technologies and machines, workers became more efficient and productive by assembly line, so called "electricity." Many innovations in manufacturing were characterized this era. At that time, design was focused on "efficiency." The industrial revolution was the transformation of the old methods of creating consumer goods into new ways of production through introduction of new technologies and machines. People were more efficient and productive. During third revolution era, people abandoned a traditional life in the countryside and moved to cities, which created a specialized and interdependent economic life and made an urban worker more dependent on the will of the employer than the rural worker had been. Design changed economic and life style with a new design notion, so called "mass production." The third era of industry came about with the advent of computers and the beginnings of automation, when robots and machines began to replace human workers on those assembly lines by "automation." Furthermore, worldwide internet, digitalization, and interactive machines emerged in people's daily life, which creates isolated individuals. Isolated individuals don't need to meet in-person to across-the-table meeting due to the fast and stable online network system. And the forth industrial revolution is in the progress of creating what has been called a "smart factory." Cyber physical systems monitor the physical processes of the factory and make decentralized decisions. Computers and automation will come together in an entirely new way; Robotics connected remotely to computer systems equipped with machine learning algorithms that can learn and control the robotics with very little input from human operators [2]. Society is already saturated with various high-end technological products. What is design for now and in the future society, and what will be able to be expected from the objects?

2. A New Era to Rethink the Notion of Design

Thus far, design has mainly been discussed in terms of the study of industrial design [3, 4] and engineering design [5]. However, society is already saturated with various high-end technological products, and design is continually reconfiguring itself from object-centered to human-centered. And current design does produce not only form or function based objects, but also produce involving what the users expect from the objects, so called *Kansei* information in *Kansei* science research in Japan. *Kansei* science was first proposed by Harada during the same period as the creation of *Kansei* engineering [6]. While *Kansei* engineering is a kind of technology that translates the customer's feeling into design specifications [7], *Kansei* science is a transdiscipline at the crossing of *Kansei* and cognitive sciences. *Kansei* science research initiated in the 1980s by Harada aimed at describing holistically users' cognitive processes related to preference and choice of products. *Kansei* science is built upon brain science, mostly cognitive neuroscience and psychophysiology, and relies on related philosophies [6]." The authors have been verified *Kansei* information aiming at evaluating individual experiences "properly" which contribute to very subjective and intuitive feelings to the objects with physiological and psychological approaches in valid. Design disciplines and research have been rapidly transforming, and not only objects but also services are target of design. Remained questions are two-fold: (1) how verify *Kansei* information (2) how to phase *Kansei* information in design to be effective? The paper is organized as follows: First, the authors review a new design discipline towards enhanced affective design, which attributes to *Kansei* information. Second, a model for understanding affective-cognitive information process is built aiming at rethinking the notion of design to propose a conceptual framework for integrating user experience into objects. The approach is to strengthen the form and function based design with aesthetically pleasing by verifying the link between product and the perceptual impacts on the users focused on affective values.

3. Multidisciplinary Approaches to Define Design

How design what the individual wants and needs? This question has fascinated researchers in design field, and continues to motivate them today. Design doesn't address a single issue anymore. Design has been more complex involving the notions of user, eco, universal, sustainable and so on. What is design, and how design for fit the new era society? "Dividing" can give an answer to.

3.1. Fundamental Aspects in Design: Perceptual vs. Conceptual

Form and functions are two fundamental aspects of design, and it is expected that the two interact to impact user's evaluations of a product. Form and function based design is mainly evaluated by the baseline knowledge or previous experience [8, 9, 10]. Typical form design is conceptualized as the specific prototype in individual memory. Thus, if the form is more typical, users can easily access to memory. In other words, more typical form leads to positive product evaluation due to its perceptual fluency. Perceptual fluency reflects the ease with which users can identify the physical identity of the stimulus on subsequent encounters, and involves the processing of physical features such as shape [11]. Also, perceptual fluency is influenced by several variables such as perceptual priming, clarification, presentation duration, or repetition [12]. Most of all, perceptual fluency is known to be enhanced through prior exposures [11]. On the other hand, conceptual fluency reflects the ease with which the target comes to individual minds and pertains to the processing of meaning which is related with semantic knowledge structures [12]. Whereas perceptual fluency is not influenced by attention or elaboration, conceptual fluency benefits from elaboration at the time of exposure [11]. In other words, perceptual fluency origins from "external feature context and subjective innate filter" whereas conceptual fluency origins from "the phase of the subjective innate filter." More details will be discussed the following section.

3.2. Verifying Kansei Values: Combine Perceptual with Conceptual

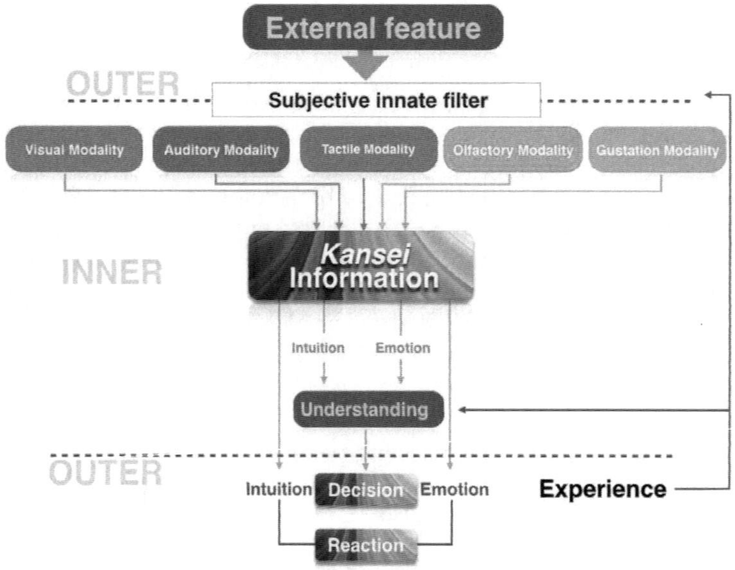

Figure 1. The relationship between *Kansei* information, reaction and decision (Kim et al., 2012; 2016)

Figure 1 represents the process of *Kansei* information and outcomes. The term of *Kansei* has been developed in Japan in order to design feelings into products, and Kim et al. [13, 14] defined *Kansei* information as Figure 1 based on the previous *Kansei* engineering and design studies. The authors postulate individual as a subjective innate filter considering the variety of subjectivity. This subjective innate filter separate individuals as "one individual." By subjective innate filter, inner and outer separated. When human captors (eyes, ears, nose...) receive a stimulus as an external feature from outer, the sensor data integrate as *Kansei* information. *Kansei* information contains full perceptive data acquired by subjective innate filter (human sensors/modalities); the information synthesizes through in the brain; and comes out as reaction such as emotion or intuition. Decision is a result of understanding. In other words, *Kansei* information as integrated sensory data is assimilated (*Kansei* process), and is sent to understanding process. *Kansei* presents as two types of outcome: emotion or intuition as *Kansei* information per se; decision as a result of understanding. The two types of outcomes influence perceptual and conceptual fluencies as familiarity, preference, aesthetic feeling, and so on; and the two interact to impact user evaluations of a new product again through the individual experience. As mentioned above, considering the definition of *Kansei*, perceptual fluency origins from "external feature condition and subjective innate filter" whereas conceptual fluency origins from "the phase of the subjective innate filter."

(a) (b)

Figure 2. Examples of perceptual fluency

Figure 2(a) presents examples of which influences perceptual fluency. While the left "fluent" are easy to read, the right ones aren't. The first letters are printed in a blurry font, and third letters featuring an upside down. These are very orthodox examples of the effects of perceptual information. Figure 2(b) presents also the example of which influences perceptual fluency. Only one green sector is different from others. The answer is no.11. If you are one of the Himba of northern Namibia, you can categorize colors and very quickly point out the standout color [15]. This example give a clear answer why some "external features" don't show the same results as others.

3.3. Integrating Affective Values to Design: Balance between Perceptual and Conceptual

In everyday lives, people are constantly making decisions from small to big, such as choosing what to wear on, and which presidential candidate to vote. Regardless of the importance of the choice, people engage in different strategies to make these decisions, by either following their feelings, intuition, or by weighting all the pros and cons. While the first processing is related to "fast thinking" which consider to be intuitive and effortless; the second processing is related to "slow thinking" which consider to be deliberative and effortful [16]. Affective values in design are influenced by the two both. Affective values have been investigated in *Kansei* researches and used as a tool to capture and convert subjective feelings about a product into concrete design parameters. Without knowing about which design gives people pleasure and satisfaction, they seek pleasant and emotional satisfaction in quality of life. Considering perceptual and conceptual information, design is a process to verify individual subjective feelings, pleasure, and satisfaction in perceptual and conceptual levels. Design researches in human-centered, universal/inclusive, UI, UX, and so on, have been focused to give individuals subjective pleasing and satisfaction. *Kansei* researches, whether science or engineering-oriented, have been focused on verifying individual subjective feelings, and having the foresight to transform design disciplines. In this paper, the authors clarify the relationship between "individual" and "object", focused on its factors to propose a new affective-cognitive approach to design.

4. The Notion for Enhanced Affective Design: *Kansei*-Oriented Affective Design

Aesthetic aspects are a potential source of pleasure for the users [17]. Aesthetic response tends to involve a number of factors. Out of those several factors, typicality has been widely researched to examine the relationship between typicality and affect. Many studies in marketing and consumer fields have demonstrated that form typicality influences product evaluation via affect. Recent previous literature shows that a typical form is usually more appealing such that there is a positive relationship between typicality and affect [18, 19]. More typical product design is more affectively pleasing [20, 21]. The right answer to the question is three-fold: what typicality represents out of the "external feature"; what typicality represents out of the "subjective innate filter"; how those typical outcomes, which comes from the external feature and from subjective innate filter, are made to meet in design.

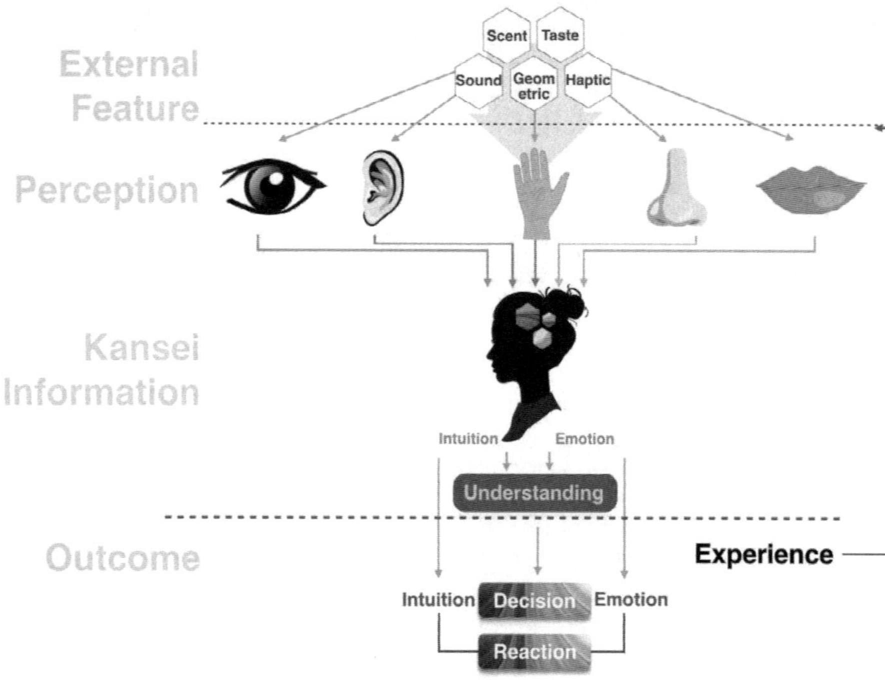

Figure 3. The relationship between perceptive data, *Kansei* information, and outcomes considering the factors of external feature (Kim&Cho, 2017)

Figure 3 shows the flow of *Kansei* information and its outcomes which evolved by the author, 2017. The most different concept from the previous definition is on "division." In the evolved-definition, not only information through "subjective innate filter," but also information of "external feature" is to be divided into factors such as scent, taste, sound, geometric, and haptic. The previous definition of *Kansei* information, perceptive information through the subjective innate filter were divided into visual, auditory, tactile, olfactory, gustation information [Figure 1]. The definition is focused on the method to verify most proper design factors, which comes from intuition, emotion, and understanding of the individual. In order to clarify more affective scent or sound, etc., and to please the users is important in the notion of design to foresight individuals' subjective satisfaction beyond the design so far. The most impressive scent, sound, and so on, is not only the best part of the experiences of the individual, but it is what the individual remembers most as the specific prototype of the design. The smell of a strawberry triggers sweet, savoy, salty and so on. The smell of a strawberry is the "fact" that provides objective information independent of an individual, whereas, the subjective feelings from the strawberry smell is dependent on the individual experience or expectation. The last definition of *Kansei* information by the authors is beneficial to affective design with the trans-discipline at the crossing of *Kansei* and cognitive science researches due to its holistic research methods. Affective design aims at human-centered, user-experience, whereas pleasure-giving design, aims at neither customized design nor personalized design.

Affective design will have to bring more values to both users/customers and designers than efforts to use/costs so that it will be a viable approach. For users, they can be provided with products with less lead time, high quality, and better preference match. For producers, they can be benefited by providing *Kansei*-oriented affective design products at a higher price with satisfied users, valid data about individual user's latent needs, and so on. It is not easy to explain what people really want and need in words. In the paper, it was reviewed what design disciplines towards enhanced affective design aiming at rethinking the notion of design to propose a conceptual framework for integrating user experience into objects that strengthen the form and function based design with aesthetically pleasing. Design disciplines have been contributing to shaping the life of human beings, as well as fostering culture and heritage. The notion of *Kansei*-oriented affective design is valid for both designers and users, because of that design disciplines and research have been rapidly transforming, and not only objects but also services are target of design. Transdisciplinary approach which was used in *Kansei* science research, is effective and sustainable in pleasing design.

References

[1] Definition of Design, http://www.icsid.org/about/about/articles31
[2] Marr B 2007 *Data Strategy: How to Profit from a World of Big Data, Analytics and the Internet of Things* (Kogan Page)
[3] Buchanan R, Margolin V 1995 *Discovering Design: Explorations in Design Studies* (University of Chicago Press)
[4] Cross N 2007 *Designerly Ways of Knowing* (Birkhäuser Basel)
[5] Pahl G et al 2007 *Engineering Design—A Systematic Approach* (Springer)
[6] Levy P 2014 Perception Theories and Kansei Design. *The proceedings of Kansei Engineering and Emotion Research 2014*
[7] Nagamachi M, Lokman A 2010 *Innovation for Kansei/affective engineering* (CRC Press)
[8] Gregan-Paxton et al 2005 When Categorization is Ambiguous: Factors that Facilitate the Use of a Multiple Category Inference Strategy. *Journal of Consumer Psychology* **15** (2) pp 127-40
[9] Hoeffler S 2003 Measuring Preferences for Really New Products. *Journal of Marketing Research* (November) pp 406-20
[10] Meyers-Levy, Tybout 1989 Schema Congruity as a Basis for Product Evaluation. *Journal of Consumer Research* **16** (1) p 39
[11] Lee A Y, Aparna A A 2004 The Effect of Conceptual and Perceptual Fluency on Brand Evaluation. *Journal of Marketing Research* **41** pp 151-65
[12] Reber R et al 2004 Processing Fluency and Aesthetic Pleasure: Is Beauty in the Perceiver's Processing Experience? *Personality and Social Psychology Review* **8** (4) pp 364-82
[13] Kim S 2012 Preference Mechanism in Product Evaluation using Automotive Image by an Approach of Kansei Information. *PhD Thesis* (The University of Tsukuba, Japan)
[14] Kim S et al 2016 Integrating Affective Values to Sustainable Behaviour focused on Kansei Engineering. *The International Journal of Sustainable Engineering* doi: 10.1080/19397038.2016.1206984
[15] http://boingboing.net/2011/08/12/how-language-affects-color-perception.html
[16] Evans J B T 2010 Intuition and reasoning: A dual-process perspective. *Psychological Inquiry: An International Journal for the Advancement of Psychological Theory* **21** (4) pp 313-26
[17] Veryzer, Jr W H 1998 The Influence of Unity and Prototypicality on Aesthetic Responses to New Product Designs *Journal of Consumer Research* **24** (4) p 374-394
[18] Carpenter G S, Nakamoto K 1989 Consumer Preference Formation and Pioneering Advantage. *Journal of Marketing Research* **26** (3) pp 285-98
[19] Folkes V S, Patrick V M 2003 The Positivity Effect in Perceptions of Services: Seen One, Seen Them All. *Journal of Consumer Research* **30** (1) pp 125-37
[20] Loken B et al 2007 Categorization Theory and Research in Consumer Psychology. *Handbook of Consumer psychology* pp 133-163
[21] Schwarz N 2004 Metacognitive Experiences in Consumer Judgment and Decision Making. *Journal of Consumer Psychology* **14** (4) pp 332-48

IOP Conf. Series: Materials Science and Engineering **234** (2017) 012026 doi:10.1088/1757-899X/234/1/012026

Kinematical Comparison Analysis on the Discus Athletes Throwing Techniques Based on Data Project

Li Junming[a], Zhou Jihe[b] and Long Ting[c]

Chengdu Sport University, China

[a]Email:605067954@qq.com; [b]email:jihezhou123@163.com; [c]email:695498130@qq.com

Abstract: In the discus final site of throwing event series game of China's track and field sport in April, 2015, three dimensional camera analytical method which is an application of kinematical data project was used on female discus athletes' discus throwing technology. And analysis was made for the top four discus throwers' last exertion action, related kinematics parameter was thus obtained. Analysis results show that: first, Lu Xiaoxin behaves better in body twist tight effect when it is left foot on the ground and in capacity of beyond devices, followed by Su Xinyue and Tan Jian, with Feng Bin relatively weaker; second, our athletes' discus shots speed is to be upgraded compared with world excellent female discus athletes; third, discus is left slightly earlier, with Tan Jian throwing in a reasonable angle, Feng Bin, Lu Xiaoxin in a larger angle, and Sue Xinyue in a smaller angle. Feng bin has a higher height of release, followed by Lu Xiaoxin and Tan jian.

1. Introduction

In recent years, China has seen a number of female discus athletes attracting the attention of the world. However, we still have a far cry from world elite discus throwers in the throwing techniques. In this paper, with the method of three-dimensional camera techniques in female discus throwers and the resolution for the last exertion movement video of top four athletes, kinematics parameters of four female discus throwers in the game are obtained. That means we can summarize movement characteristics and shortcomings of each person through comparative analysis in a way to throw a light on theoretical basis and technical reference for the improvement of the Chinese women's discus throwing techniques.

2. Subjects and Methods

2.1. Subjects

The subjects in this study were top four female discus throwers in the discus final of throwing event series game of China's track and field sport in April, 2015. According to the scores, the top four were Lu Xiaoxin, Tan Jian, Su Xinyue, and Feng Bin (Lu, Tan, Su, Feng for short). We will introduce some basic information about these four athletes in Table 1.

Table 1. Subject Information and Trials Used for Analysis

Name	Standing Height（cm）	Personal Record （m）	Single Support （m）
Lu Xiaoxin	177	61.47	61.47
Tan Jian	175	64.44	61.32
Su Xinyue	178	63.90	61.00
Feng Bin	181	60.75	60.75

2.2.Methods

Three dimensional camera analytical method was mainly used. In the discus final of throwing event series game of China's track and field sport in April, 2015,three dimensional fixed-point high speed cameras were used to record the movement of the top four female discus throwers. Two JVC GC-PX10AC cameras were placed on the rear and right side around the discus circle with an approximately 90°angle at a shooting frequency of 50Hz. SignaltEC3D video analysis software was applied to parse the video and Matsui Xiuji model of human body (21 points, 16 links) was selected for experimental use. Original data was handled through low-pass filter smoothly, at a cut-off frequency of 8Hz.

3. Results and Analysis

Last exertion stage of throwing discus is the movement process from the time when discus throwers' left foot is on the ground to the moments of discus shots. Therefore, the last exertion stage has a decisive role on discus results[1], which is targeted on back-to-back rotating action. The last exertion technique makes full use of rotating momentum and favorable release posture, thus exerting all the body strength to the discus through the arm and hand throwing. As a result, maximum release speed and optimum angle of release can be obtained, reaching the best effect of throwing and getting excellent results[2].

3.1. Time analysis on the last exertion stage

We can know from Table 2 that the time used in the last exertion stage for Lu, Tan, Su, Feng is 0.18 second, 0.12second, 0.14second, and 0.24second respectively. While the time for world elite female discus throwers is between 0.1 and 0.2 second[3], we can see that three of our athletes have a reasonable time control except Feng with longer time.

Table 2. Time Used in the the Last Exertion Stage

Name	Lu Xiaoxin	Tan Jian	Su Xinyue	Feng Bin	World elite Female Discus Throwers
Time（s）	0.18	0.12	0.14	0.24	0.1-0.2

3.2.Analysis of kinematic parameters on the moment of left foot on the ground

From Table 3, on the moment of left foot on the ground, the pull angle of Lu, Tan, Su and Feng is respectively 121.8 °, 129 .0°, 125.0 °, and 142.1 °. These data indicate that Lu keeps a stronger capacity in beyond devices when it is the moment of left foot on the ground and has a better control in keeping the discus behind compared with the three others, with Su and Tan followed, and Feng the weakest.

From Table 3, on the moment of left foot on the ground, the shoulder-hip angle of Lu, Tan, Su, and Feng is 65.3 °, 60.2 °, 61.6 °, 50.4 ° respectively. We can conclude from the data of shoulder-hip angle that among four athletes, Lu has a better torso twist pose and a posture of hip well beyond the shoulder, followed by Su and Tan, with Feng the worst.

Speed difference of discus release between the left and right shoulder can be considered as an evaluation index for the discus speed kept to a certain degree in the last exertion stage[4]. From Table 3, speed difference of discus release between the left and right shoulder of four athletes is respectively 3.27m/s, 2.42m/s, 2.52m/s, and 2.07m/s. Lu has a larger shoulder speed difference compared with the three others, which indicates Lu has a larger range of opening among chest, shoulder and arms, and stronger capacity in beyond devices, followed by Su and Tan. Feng has a relatively smaller range of opening among chest, shoulder and arms, and weaker capacity in beyond devices, thus effecting the last exertion.

Table 3. Kinematics Parameter on the Moment of Left Foot on the Ground

Name	Lu Xiaoxin	Tan Jian	Su Xinyue	Feng Bin
Shoulder-Hip Angle（°）	65.3	60.2	61.6	50.4
Pull Angle（°）	121.8	129.0	125.0	142.1
Shoulder Speed Difference（m/s）	3.27	2.42	2.52	2.07

3.3. Analysis on the speed of discus release

The speed, angle and height of discus release is key factors influencing the throwing distance. The study has showed that only the speed of release is the most significant among three factors[5]. We can know from Table 4 that the speed of four athletes is 23.65m/s, 23.09m/s, 23.24m/s, 22.57m/s in comparison to 25m/s of world counterparts[3], which means improvement for our discus throwing speed. Of four throwers, Feng performs the slowest in the speed of discus release in dire need of strengthening special training.

Table 4. Kinematics Parameter on the Moment of Discus Release

Name	Lu Xiaoxin	Tan Jian	Su Xinyue	Feng Bin	World elite Female Discus Throwers
Speed of Release（m/s）	23.65	23.09	23.24	22.57	About25

3.4. Analysis on the relevant angle of discus release

From the Table 5, the angle of release for four discus throwers is 41.2°,36.4°,32.8°, and 41.7°respectively. From some documents, we can know that the reasonable range of release angle for female discus throwers is between 35°and 40°[4]. Among four, Tan behaves best with a reasonable release angle, while Feng and Lu behave with a larger angle and Su with a smaller angle.

In the last exertion process, shoulder-hip angle should gradually decrease, and pull angle should gradually increase, thus the stored energy in the chest, shoulders and arms can be released, at the same time, the right foot should actively pedal with the left foot forming the stable support on the left, enabling the shoulder axis to rapidly surpass the hip axis, which will lead the rapid throwing of discus[3]. We can know from Table 5 that the shoulder-hip angle of Lu, Tan, Su, and Feng in the moment of discus shot is respectively 18.1 °, 18.5 °, 17.5 °, and 10.4 °, which shows the right shoulders of four athletes all fall behind their right hips in the moment of discus shot. However, the most optimum shoulder-hip angle is about 0 °according to some related literature[4]. Therefore we can see that Feng has a larger force range, with Lu, Su, and Tan relatively weaker. Once the last exertion range is smaller, the force produced by body twist fails to fully pass to the discus. Pull angle of discus release for four female throwers is 167.5 °, 163.3 °, 170.0 °, and 170.9°respectively. That means that four players fail to fully unleash the elastic potential energy reserved in their shoulders, hips and arms in a result of throwing the discus slightly earlier.

Table 5. Parameter on the Relevant Angle of Discus Release

Name	Lu Xiaoxin	Tan Jian	Su Xinyue	Feng Bin
Shoulder-Hip Angle（°）	18.1	18.5	17.5	10.4
Pull Angle（°）	167.5	163.3	170.0	170.9
Angle of Release（°）	41.2	36.4	32.8	41.7

3.5. Analysis on the height of discus release
The height of discus release is closely associated with technique characteristics and the height of athletes[5]. From the Table 1, the height of four discus throwers is 1.77m, 1.75m, 1.78m, 1.81m. From the Table 6, the height of discus release of four female throwers is 1.607m, 1.561m, 1.674m, 1.712m respectively. The gravity height on the moment of discus release is 1.090m, 0.995m, 1.030m, 1.130m respectively. Through the comparison analysis of data above, we know that Feng and Su have a higher release, followed by Lu, with Tan the lowest.

Table 6. Kinematics Parameter on the Moment of Discus Release

Name	Lu Xiaoxin	Tan Jian	Su Xinyue	Feng Bin
Height of Release（m）	1.607	1.561	1.674	1.712
Gravity Height（m）	1.090	0.995	1.030	1.130

4. Conclusions
We can draw the following conclusions from the kinematical comparison analysis on the discus athletes' last exertion techniques based on data project.

4.1. The final force phase is used to conclude
Feng uses a larger time of 0.24s in the last exertion stage with the three others relatively reasonable in the time control.

4.2. Left -footed momentary kinematic conclusion
According to the data of pull angle of four athletes on the moment of left foot on the ground, Lu keeps a stronger capacity in beyond devices at a pull angle of 121.8 °and has a better control in keeping the discus behind compared with the three others, with Su and Tan followed, and Feng the weakest. According to the data of shoulder-hip angle, Lu has a better torso twist pose and a posture of hip well

beyond the shoulder at a should-hip angle of 65.3°, followed by Su and Tan, with Feng the worst. According to the shoulder speed difference, Lu has a larger range of opening among chest, shoulder and arms at a speed difference of 3.27m/s, and stronger capacity in beyond devices, followed by Su and Tan. Feng has a relatively smaller range of opening among chest, shoulder and arms, and weaker capacity in beyond devices, thus effecting the last exertion.

4.3. Discus shot speed conclusion

Lu and Su have a relatively faster speed of discus release at 23.65m/s and 23.24m/s respectively. However, improvement is still needed compared with world elite female discus throwers. Of four throwers, Feng performs the slowest in the speed of discus release of 22.57m/s in dire need of strengthening special training.

4.4. The conclusion from the Angle of the discus

Tan behaves best with a reasonable release angle of 36.4°, while Feng and Lu behave with a larger angle. If Su increases her release angle better results can be scored. The data of pull angle show that four players fail to fully unleash the elastic potential energy reserved in their shoulders, hips and arms in a result of throwing the discus slightly earlier.

4.5. Discus shot a high degree of conclusion

The comparison of data about release height show that Feng and Su have a higher release, followed by Lu, with Tan the lowest.

Acknowledgements

First, I would like to extend my sincere gratitude to the cultivation and education of Chengdu Sport University. Second, high tribute shall be paid to my supervisor, Mr.Zhou, for his instructive advice and useful suggestions on my thesis. I am deeply grateful of his help in the completion of this thesis. Special thanks to my classmats and friends for their supports and help for my thesis.

References

[1] Ji Peng, Liang Dongxu, Analysis on the left side supporting technology of discus final exertion, Journal of Handan Polytechnic College, 2003, 03:77-78.

[2] Su Qiao, Kinematical Analysis of Chinese elite discus thrower Nuermaimaiti·tulake, Chengdu Sport University, 2014.

[3] Wang Dong, Kinematical Analysis on spatial and temporal characteristics of top eight females' discus throwing technique in the Eleventh National Games, Capital University of Physical Education and Sports, 2011.

[4] He Yunshan, Kinematical Analysis on the last exertion technology of female discus throwing in the Eleventh National Games, Shanxi University, 2011.

[5] Zhou Xuebing, Chen Dong, Kinematical Analysis on Wu Jian's discus throwing technology, Journal of Shandong Sport University, 2006, 02:69-71+75.

IOP Conf. Series: Materials Science and Engineering **234** (2017) 012027 doi:10.1088/1757-899X/234/1/012027

The Study to the Theory of Human-Machine-Integration in Product Design

Min Gao and Renhe Zhou

Art&Design college, Huaiyin Institute of Technology, Jiangsu Province, China

hyitzhourenhe@163.com

Abstract. The paper puts forward a design concept named Human-Machine-Integration which refers to the shape of the product designed according to the location and contact type when the product is used. It considers the double properties of function and aesthetics to product and makes the designer combine the formal principle with Human-Machine Engineering theory together to mould the product shape conforming to the easy-to-user principle with the consideration of form beauty at the same time. It regards the man - machine – environment as a whole body and has practical significance to the system design of the product. When a product is designed according to Human-Machine-Integration theory, contact parts form and contact area and the harmonious relationship between forms of contact parts and overall product should be considered. And the Convex-concave match of product part and body part should be complement with each other and tolerant match of product part is contacted with the human body part. At the same time, the purpose of form should be fully considered. Correspondence, Integrity, Systematicness and Independence are four characteristics of Human-Machine-Integration design.

1. Introduction

The generation of artificiality accompanied with the creation and extension of the form. The process of shape extension is also the process of searching for good shape[1]. In this course, the form is gradually combined with the user and use status according with the development of Human-Machine Engineering. And it shows the practical value besides the aesthetic value[2]. The first property of form is materiality determinate that its function to the objects is moulding. The double properties of function and aesthetics to product make the designer combine the formal principle with Human-Machine Engineering theory to mould the product shape conforming to the easy-to-user principle with the consideration of form beauty at the same time.

Form-Fit means close match to others between shapes[3]. The Form-Fit design is finding the corresponding relationship between shapes such as up and down, left and right, positive and negative according to the basic function requirement among shapes.

It is from the collinear and conformal theory in Graphics. That means two shaped have common edge line when shapes are combined[4]. The method of shared shape reflects the meaning of Form-Fit[5]. As to the 3D product's performance, collinear and conformal became coplanar and shared body. Corresponding relationships between each other make products or different parts of product to join each other and become the new whole one. The research object of this article is Human-Machine-Integration theory which refers to the shape of the product designed according to the location and contact type when used. The product form is designed to considerate the corresponding relationship

IOP Conf. Series: Materials Science and Engineering **234** (2017) 012027 doi:10.1088/1757-899X/234/1/012027

between touch location and operating type such as pressing ,gripping , dialing , pulling , treading , sitting. Then the form of product is designed to adapt to the human body and make the product form to convey semantic and guide function.

2. The type of Human-Machine-Integration

According to the corresponding relationship with each other, Human-Machine-Integration can be divided into surface correspondence and body correspondence. When the operation location to the body is one surface, the product surface is designed to be a complementary form with the contact surface. The design of various buttons is the most common example. Body correspondence means one part of the body and part of the product or a product that is both inclusive or accommodate. The whole or part of product can be held by one part of body, or the product has certain instruction space to accommodate certain part of body. This indicative modelling language has semantic role. Human-Machine-Integration can also be divided into internal and external correspondence according to the space position relations. Internal correspondence refers to the internal part of product can accommodate certain part of the body. While external correspondence refers to the operation area is located outside of the product.

For example as Fig.1 shows in the picture, computer's button is the part with surface contact is very frequent. So the keys are designed as slightly concave shape to make the user to put fingers forward the trend of completion. Left and right surfaces of the mouse can also be seen as the representative to match each finger with product as Fig.2 shows. The cavity in two sides can accommodate for the thumb, ring finger and little finger. Also the model makes the mouse to be an aesthetic shape. Hand shape should be fully considered when design handheld tools.

Figure 1. This figure shows the design of computer keyboard. The keyboard is designed based on the shape of human finger. And it indicates the method of application with its shape.

Figure 2. Mouse is the product that the designer should take more ergonomics factor into account. Frequent operation makes its shape must cater human hand.

3. The significance of Human-Machine-Integration form

According to the Gestalt theory on psychology, an open graphic has the trend of completing good form. People will close or fill a graphic vacancy to complete a whole body. When the surface or body of product has open property, their own gestalt trend will make users to fill the vacancy so as to achieve the effect of man-machine integration.. Product forms act as the instruction to indicate the operation behaviour of people and prevent improper operation aiming at more reasonable, safer and more comfortable operation. Direct experience and indirect experience work together, through a series of reinforcement, to form fixed good use habits and form the correct behaviour patterns. Finally the purpose of the comfortable and safe operation can be achieved.

Let's come to the model evolution of Video Game Controller. Controller is the main place of human-computer interaction for operators to operate games. It is controlled by the operator's hands. Fig.3 shows that the initial shape of Video Game Controller is a rectangle, without considering the hand's interaction with the controller. With the human comfort research, the controller's appearance has changed a lot to prominent dual handle. It is suitable for both hands gradually and it is becoming more and more humanized as the new type controller in Fig.4.

Figure 3. The model evolution of Video Game Controller.

Figure 4. This is the new type Video Game Controller which is suitable for both hands. Based on Human-Machine-Integration theory, the controller is designed to fit the shape of hand and gain the success of performance.

4. Design key points and main characteristics

Human-Machine-Integration theory is about the product shape fits the human body, and it is a form design pattern combined form beauty and function value. The research of Human-Machine-Integration theory will help us broaden our design vision and strengthen our ability to create new form. The product's specific function and operating status must be deeply understood when track to the Human-Machine-Integration design. And the basis point is the interaction between the user and the product.

4.1. Design key points

Human-Machine-Integration form design's key points are as follows:

- The designer must comprehensively consider contact parts form and contact area and the harmonious relationship between forms of contact parts and overall product. Then the model of product is finally determined. The system is composed of this three. Any one's change will affect the other two parties and will ultimately affect the Human-Machine-Integration form of the product.
- The Human-Machine-Integration form include the Convex-concave match of product part and body part to complement each other and tolerance match of product part is contacted with the human body part to accommodate each other. The type of Human-Machine-Integration should be ensured according to the different operating state. Then the corresponding design is carried on.
- When design the overall form based on Human-Machine-Integration theory , the purpose of form should be fully considered. The designer needs to combine specific function with the corresponding aesthetic form to make a systematic design.

4.2. Main characteristics
Therefore, the main characteristics of Human-Machine-Integration design are listed as the following: Correspondence, Integrity, Systematicness and Independence.

- Correspondence can also be called as' coordination. It is refers to the corresponding relationship to each other between human and machine such as concavo-convex relationship, embedded relationship, tolerance relationship, etc. This is the most basic morphological characteristics of Human-Machine-Integration. A product with good Human-Machine-Integration form has good human-machine interaction relationship and high cooperation degree.
- Integrity includes two aspects. The one is that as a product with good Human-Machine-Integration design, the conjunction part and contact parts of users should correspond to each other to get a whole. At the same time conjunction part is a part of the entire product, and the products including conjunction parts is a whole.
- Systematicness means Human-Machine-Integration design is not the design that separately considers function value, but the synthesis combined with function value and aesthetic value. It's a systematic design.
- Independence refers to the fit part has its own function and form characteristics, so it is relatively independent individual. Finally all individuals are unified in the whole product.

5. Design cases
Product is created to improve the human life. Design from the perspective of Human-Machine-Integration can guide the design of industrial products and seek for solutions and design ides for products. It's a new product design thought to look for embodiment of Human-Machine-Integration in various sorts of modelling form and interaction of product and user and to apply into practical design. There are some practical design cases by my students as follow in Fig.5, Fig. 6 and Fig. 7.

IOP Conf. Series: Materials Science and Engineering **234** (2017) 012027 doi:10.1088/1757-899X/234/1/012027

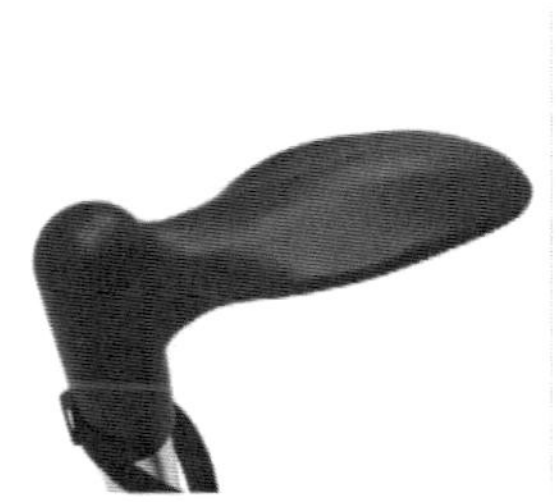

Figure 5. The picture shows the design of walking aid 's handrail. Walking aid is designed for disadvantaged groups to help them walk. The handrail is the part that the hand holds the walking aid. According to the Human-Machine-Integration theory, the shape of handrail is designed as picture shows.

Figure6. The design is for fingerprint identifier.The user put his or her thumb on the interactive surface of fingerprint identifier ,and then the identifier give the result. The place where the thumb tontacts the interactive surface is designed with correspondence to the shape of thumb to give a sign to use the machine.

Figure 7. Child amblyscope is the product that cures children amblyopia which is more and more common among children nowadays. The part that put the face on is designed in cooperation with child's face and eyes. So it is easy for child to judge how to find the place before the medical instrument.

6. Appendices

Human-Machine-Integration design concept fully considers the function value and use condition of the product. It regards the man - machine – environment as a whole body instead of only considering the product itself. The traditional Chinese design thought that implement is used by people and it is an extension of the human body also embodies the design concept of Human-Machine-Integration. Form is the carrier of function. Human-Machine-Integration design concept extents the morphological function to the operation process of product and deepen the function of the form. So it has certain practical significance to the system design of the product.

Acknowledgments
The authors gratefully thank for the students of Kong Xiaofang, Ge Lixia and Wang Xiaoyun for they provide their design works as Human-Machine-Integration cases in this article. The second author Renhe Zhou gave many useful suggestions to this paper and helped to finish it. The article is funded by [1] *Talent Lifted Project* held by Jiangsu Science and Technology Association[2]Project of *International cooperation in science and technology* in Huaian City(*HAC201616)*.Thanks to all the people who gave their selfless help to this article.

References
[1] Zhang Wen 2007 *Product form fit design research*（Nanjing：Journal of Nanjing University of the Arts）p146.
[2] Luo Qi 2011 *Matching Shape and Product Design (1994-2011China Academic Journal Electronic Publishing House)* p98.
[3] Liu Guoyu and Shen Jie 2000 *The basis form of product design* (Beijing: China Light Industry Press) p5.
[4] Sang Ruijuan 2010 *The functional Value in Form Conjunction of Furniture* (Nanjing: Furniture and Interior Decoration)p60.
[5] Zhang Fan 2012 *A preliminary exploration of packaging design (*Beijing: Journal of Beijing Technology and Business University)p69.

Institute of Physics
Dirac House, Temple Back
Bristol BS1 6BE UK

ISSN: 1757-8981
ISBN 978-1-5108-4777-4

Human Factors 2024

Wageningen, The Netherlands
8-9 October 2024

The Royal Institution of
Naval Architects